Lion

Animal
Series editor: Jonathan Burt

Already published

Crow
Boria Sax

Fox
Martin Wallen

Ape
John Sorenson

Ant
Charlotte Sleigh

Fly
Steven Connor

Owl
Desmond Morris

Tortoise
Peter Young

Cat
Katharine M. Rogers

Snail
Peter Williams

Cockroach
Marion Copeland

Peacock
Christine E. Jackson

Hare
Simon Carnell

Dog
Susan McHugh

Cow
Hannah Velten

Penguin
Stephen Martin

Oyster
Rebecca Stott

Swan
Peter Young

Bear
Robert E. Bieder

Shark
Dean Crawford

Bee
Claire Preston

Elephant
Daniel Wylie

Rat
Jonathan Burt

Moose
Kevin Jackson

Snake
Drake Stutesman

Eel
Richard Schweid

Falcon
Helen Macdonald

Duck
Victoria de Rijke

Whale
Joe Roman

Hare
Simon Carnell

Parrot
Paul Carter

Rhinoceros
Kelly Enright

Tiger
Susie Green

Pigeon
Barbara Allen

Salmon
Peter Coates

Camel
Robert Irwin

Lion

Deirdre Jackson

REAKTION BOOKS

To Daniel

Published by
REAKTION BOOKS LTD
33 Great Sutton Street
London EC1V ODX, UK
www.reaktionbooks.co.uk

First published 2010
Copyright © Deirdre Jackson 2010

Publication of this book has been made possible by a grant from the
Scouloudi Foundation in association with the Institute of Historical
Research

All rights reserved

No part of this publication may be reproduced, stored in a retrieval
system or transmitted, in any form or by any means, electronic,
mechanical, photocopying, recording or otherwise without the prior
permission of the publishers.

Printed and bound in China by Eurasia

British Library Cataloguing in Publication Data
Jackson, Deirdre Elizabeth, 1965–
    Lion. – (Animal)
    1. Lions. 2. Lions – Mythology.
    3. Lions in art. 4. Lions in literature
    I. Title II. Series
    599.7'57-DC22

ISBN: 978 1 86189 655 1

Contents

Preface

Majestic, noble, brave – the lion is the King of Clichés. Judging from the wealth of lore and legend accumulated over millennia, lions have occupied a greater place in peoples' imaginations than any other animal. During the course of time, people and lions crossed paths; sometimes the outcome was happy, sometimes tragic, but it was generally memorable. Our preoccupation with the predator is reflected in a range of different cultures; the lion is even revered in places where it has never been indigenous. Tracking the lion is a monumental task. This is not, as the writers of the medieval bestiaries would have us believe, because it erases its spoor with its tail, but because its tracks are all too plentiful and lead from the dark forests of northern Europe across the Mesopotamian plains to the imperial palaces of China.

Although the lion is not the largest or fastest felid, its position as King of Beasts has rarely been challenged. People have been captivated by lions since Palaeolithic times and the animal, with its tawny pelage and luminous eyes, continues to beguile us. While we encroach on the shrinking habitats of the last remaining lions on earth, showing an alarming disregard for their future, our fascination with legendary lions persists. *The Lion King* (Walt Disney, 1994), screened worldwide, earned a staggering us$783.8 million at the box office.

Despite the animal's high profile, it remains enigmatic. We know little about the earliest lions and their ancestors, and several aspects of the lion's form and function remain to be explained, including the precise purpose of the mane. Moreover, hours spent by scientists observing the animal have not answered every question concerning their behaviour, including why they often hunt in groups and lead more complex social lives than any other big cat. These issues are important because the more we learn about lions, the better we can create strategies for their conservation.

In historic times, the lion could be found in Europe, south-west Asia, and throughout Africa, but its distributional range has been drastically reduced. In the last two hundred years alone, lions have been eradicated from Syria, Iraq, Iran, North Africa, much of South Africa, Pakistan and almost all of India. The vulnerable African lion (*Panthera leo*) is now confined to the sub-Sahara, and its critically endangered Asian cousin (*P. leo persica*) is found only in the Gir Sanctuary, Gujarat State, India, home to an inbred population of about 300 animals. Few books in English are devoted to the Asiatic lion, and many people remain unaware of its existence, let alone its struggle to survive. Lions are threatened by poachers, hunters and diseases, but above all by the large-scale destruction of their habitats. Preventing the complete disappearance of these exceptional animals means allowing them to share our world, and giving them room to roam.

Introduction

The early history of the lion is obscure, but one aspect is clear: people first crossed paths with the predator millions of years ago. The jawbone of a lion-like cat, active 3.5 million years ago, was discovered in Laetoli, Tanzania, where Mary Leakey and her co-workers uncovered hominid footprints of comparable antiquity. To the north at Olduvai Gorge, one of the most important hominid fossil sites in the world, excavations revealed the first definitive fossil of a lion that lived nearly 2 million years ago.[1] More widespread than any other large terrestrial mammal apart from humans, these archaic animals were the distant ancestors of the modern lion (*Panthera leo*).

Lions spread from East Africa into Asia and Europe then travelled to the New World, advancing as far south as Peru. Fossils found in Italy show that they arrived in the region around 600,000 years ago. 'The appearance of lions in Europe', writes biologist Bruce Patterson, 'roughly coincides with the first appearance of hominids in the temperate zone, dispersing from the tropics. Lions, leopards, spotted hyenas, and humans all moved in parallel at roughly the same time, suggesting a common environmental trigger.'[2]

Few people who hurry past the bronze lions at the base of Nelson's Column are aware that the fossilized toe bone of an actual lion was discovered on the south side of Trafalgar Square

Almost two million years ago, lions roamed Olduvai Gorge, Tanzania.

in 1957. Deposited by the River Thames, the fossil proved that around 125,000 years ago, during a relatively balmy respite between Ice Ages, lions lived in what is now central London, sharing the lush river valley with hippopotamuses, narrow-nosed rhinoceroses, straight-tusked elephants and fallow deer.[3] No people were present in Britain during this interglacial period, leaving the lion with little competition for hippo steaks.

Overcome by an irresistible urge to see the toe of the Trafalgar lion, I make my way to the Natural History Museum in South Kensington. Soon I find myself at the back of a very long queue of little people. It is half term and every child in Britain has assembled here as part of a great migratory herd. The fossil is not on permanent public display, so my genial guide, curator Andy Currant, leads me to a vast storeroom, a prehistoric charnel house filled with bones of every conceivable size and shape, and some that beggar belief. The tiny toe bone may not seem as impressive as a gigantic rhino skull in a nearby cabinet but, as I look at it, the traffic and squealing hordes fade into the distance and a lion dashes into view, a great cat that gives me a sidelong glance and races down the corridor.

For palaeontologists like Currant, the Trafalgar toe bone is relatively modern. Jawbones, teeth, femurs, skulls and countless other lion fossils in the Natural History Museum, London, and other collections around the world, are hundreds of thousands of years older. Lion fossils discovered in the small seaside town of Pakefield, Suffolk, for example, date back 700,000 years. Judging from the remains of giant beavers, hyenas, mammoths, hippos, giant deer, bear, wolves, sabretooth cats and other mammals found on site, the lions had plenty of company. The remarkable find of flint tools, first discovered at Pakefield in 2000, show that humans also lived there around 700,000 years ago, a fact that has caused scientists to push back by 200,000 years the date of the earliest human presence in Britain.[4] We can only speculate how early humans coped with large carnivores, which would have presented competition for resources and posed a physical threat. One thing is certain, however: Pleistocene lions were successful in adapting to a wide range of climates and habitats, which is why lion fossils are found in locations as far apart as Africa and Alaska. 'With the exception of modern humans, no other large land mammal has ever had a comparable range.'[5] The earliest known paintings discovered anywhere to date supplement the fossil evidence and provide additional proof of the lion's presence in Europe.

On 18 December 1994, Jean-Marie Chauvet, accompanied by two friends, was exploring the limestone cliffs facing the Ardèche River in southeastern France when he discovered a cave, almost 2,000 feet long, consisting of five linked chambers.

Lions on the walls
of Chauvet cave
in southeastern
France, c. 30,000
–32,000 BCE.

As the trio stepped gingerly across the floor of the cave, trying to avoid crunching underfoot the bones and teeth of cave bears that had lain undisturbed for millennia, the light from their lamps illuminated a mammoth depicted on the ceiling. Further exploration revealed that the cave was covered with hundreds of paintings of animals. Horses, bison, deer, aurochs, ibex, mammoths and bears lined the interior alongside rare pictures of rhinoceroses, an owl, the earliest leopard (*Panthera pardus fossilis*) in Palaeolithic art, and over 70 lions – far more than at any other prehistoric site.

Chauvet and his companions were able to distinguish the forms of individual felines but, most spectacularly, in the dark recesses of the last chamber, they encountered an entire pride massed together in a monumental mural without precedent or parallel. Radiocarbon tests revealed the wall paintings to be at least 32,000 years old, making these the oldest paintings in existence (15,000 years older, for example, than the famous murals at Lascaux.)

In Chauvet cave (named after its discoverer) lions engage in typical behaviour. Several hunt together, one lion bares its teeth and another rubs against its companion. Some lions have tail tufts and several have black spots on their muzzles near the base of their whiskers, an accurate detail that could only have been observed by someone who had seen a lion at close range. Clearly, the people who painted these images were familiar with lions, and the animals tolerated their presence. The paintings are not easily interpreted, but they attest to the early co-existence of man and beast.

Since fossil records do not provide evidence of soft tissue or hair, the paintings of Chauvet are an invaluable record of the appearance of Pleistocene lions. Curiously, none of them have manes, not even a male whose scrotum is clearly visible. This suggests that these archaic European lions had sparse manes or none at all. Fossils found in abundance in many different Eurasian sites indicate that these animals were larger than modern lions, were able to adapt to many different habitats and climatic conditions, and preyed on a wide variety of animals including horses, bison, deer, ibex and aurochs. Chauvet cave reminds us that people and lions once lived in close proximity in an untamed European landscape, a scenario that is difficult to imagine today while gazing at the well-tended vineyards of the Ardèche.[6]

In Germany archaeological excavations have revealed further artistic evidence of man's long-standing relationship with lions. In 2002 archaeologists working at Hohle Fels Cave near Ulm uncovered a tiny human figure with a lion's head carved from mammoth ivory. Tests showed that it dated from *c.* 31–33,000 years ago, making it one of the earliest figural sculptures in existence.[7] The ivory statuette, which stands about as tall as a toothpick, was the source of particular interest since a similar, but considerably larger, lion-man had been discovered in 1939 at a neighbouring site.

The larger lion-man, estimated to be 31–32,000 years old, was found in a fragmentary state in a cave at Hohlenstein-Stadel on the eve of the Second World War. Consigned to a cigar box, the jumbled assortment of almost 200 tiny pieces of mammoth ivory was left untouched until 1969 when an archaeologist painstakingly pieced them together. A child playing in the same cave in the mid-1970s chanced upon a piece of the ivory muzzle and it slid into place as easily as Cinderella's slipper. Mysterious hybrids, these carvings, which fuse felid and human characteristics, confirm that lions were part of our distant ancestors' mythological landscape as well as their physical environment.

Having spread throughout Europe, lions padded across the land bridge that once spanned the Bering Strait, and settled in Beringia, a relatively ice-free region of grassy steppes, which included parts of present-day Siberia, Alaska and the Yukon. Images of lions basking in the sun are so familiar that the concept of Ice Age lions requires a conceptual leap. But during the last Ice Age, Beringia attracted herbivores and served as a larder for lions who showed up on the scene around 300,000 years ago. Beringian lions feasted on wild horses, caribou, musk oxen and bison, and the odd baby mammoth. In the summer of 1979 a freeze-dried bison that had lived near Fairbanks, Alaska, some 36,000 years ago was discovered during mining operations. Christened 'Blue Babe' in honour of Paul Bunyan's legendary ox, the animal was extremely well preserved. Exceptional climatic conditions had created extraordinary evidence.

When the bison mummy was examined, a sorry tale emerged. Its muzzle was marked by puncture wounds that could only have been inflicted by a lion. Large scratches on the animal's hind legs and rump showed that the lion that bit Blue Babe's nose had one or two accomplices who tried to bring down the heavy herbivore by mounting a rear assault. The lions had savoured mouthfuls now missing from the mummy, but the greater part of the bison was preserved when temperatures plummeted and the carcass froze. One of the lions shattered its tooth, leaving a fragment embedded in the bison's neck. This crucial piece of forensic evidence confirmed that lions had killed Blue Babe. Permafrost had protected the bison's body to such an extent that congealed blood was visible at the base of the wounds, the bones retained marrow, and the skin a layer of fat.[8]

Ivory statuette of a 'Lion Man', carved approximately 32,000 years ago, found in pieces in a cave in Hohlenstein-Stadel, Baden-Württemberg, Germany.

Scientists apply the label *Panthera atrox* ('fearsome or cruel lion') to American lions of the Pleistocene, like the ones that hunted Blue Babe. At first these animals were unable to cross the ice sheets that blanketed the North American continent, but when the ice receded they headed south, travelling as far as South America. Many fossils of *P. atrox* and other big cats have been discovered at the famous Rancho La Brea tar pits in Los Angeles. Most of the lions died in pursuit of animals that were trapped in the natural deposits of asphalt, but some may have inadvertently stepped into the pools and come to a sticky end.[9] At the nearby junction of Wilshire Boulevard and Fairfax, in 2006, scientists uncovered the remains of hundreds more Ice Age mammals, including a lion skull. Judging from their bulky bones the lions of the New World were awesome predators, weighing considerably more than modern ones.

Eurasian lions can be divided into two groups: *P. l. fossilis*, which flourished *c.* 500,000 years before the present, and its descendant, *P. l. spelaea* ('cave lion'), which thrived between 300,000 and 10,000 years ago. The latter, portrayed on the walls of Chauvet cave, were probably 8 to 10 per cent bigger than their modern counterparts and, as stated above, had sparse manes or none at all. Recent studies have revealed that the Pleistocene lions of Eurasia and America are closely related in genetic terms and share common skeletal features.[10] Both groups became extinct at the end of the epoch, around 10,000 years ago. Explanations for their disappearance vary. Climate change probably played a part, as did the expansion of forests, the reduction of grassy steppes and the subsequent decline in numbers of large herbivores on which the lions preyed. Were these primitive lions the direct ancestors of modern ones? It was once thought that all lions were closely related, but genetic tests suggest that modern lions may have descended from a single population that developed independently in sub-Saharan Africa around 320,000–190,000 years ago.[11]

The first major fork in the modern lions' family tree represents a split between African and Asian lions that is believed by some scientists to have occurred 100,000 years ago with the emergence of the Asian subspecies *Panthera leo persica*. Asian lions differ slightly in appearance from their African cousins. Most have a fold of skin running along their underbellies and the tufts of hair on their tails and elbows are often more prominent. In addition, the skulls of Asian lions have two small apertures, instead of one, to allow nerves and blood vessels to connect with the eyes. Finally, males have sparser manes than their African counterparts. DNA drawn from African and Asian lions confirms that the two types are genetically as well as geographically distinct. Once ranging from Greece to the Indian subcontinent, with abundant populations in the Near and Middle East, the critically endangered Asian lion is now confined to the Gir Forest National Park and Wildlife Sanctuary,

Asiatic lions (*Panthera leo persica*).

Gujarat state, western India, where an estimated 300 lions survive in a territory reduced to 1,450 square km (560 square miles).

Isolated and inbred, Asian lions are almost identical to one another in genetic terms. Stephen J. O'Brien, the scientist who made this discovery, explains, 'Serengeti and even Ngorongoro crater lions showed rich diversity in DNA fingerprint patterns, but the Gir lions were all identical. It was as if they were all clones or identical twins. This was the most genetically uniform population we had ever observed.'[12] Further research revealed that the Gir lions were separated from neighbouring Asian prides when the Kathiawar peninsula was cut off from the mainland around 2,500 years ago. The distinctive physical characteristics of the Asian lions of Gir, originally thought to be the result of adaptations or modifications developed thousands of years ago, were revealed to be relatively recent signs of severe inbreeding. The implications of this study are sobering. Surviving Asian male lions have low levels of testosterone, high incidences of abnormal sperm and associated fertility problems. In addition the genetic homogeneity of the animals places them at risk from diseases, which could rapidly spread from one lion to the next.

Lions belong to the Kingdom: Animalia; Phylum: Chordata (Vertebrates); Class: Mammalia; Order: Carnivora; Family: Felidae; Subfamily: Pantherinae; Genus: *Panthera* and Species: *Panthera leo*. Some scientists have suggested splitting lions into as many as 24 different subspecies and others have argued that all lions should be grouped together. Taxonomy is a contentious issue because even closely related lions can differ markedly in appearance, and lions exhibit greater sexual dimorphism than any other felid (males are differentiated from females by both their size and their manes). Frederick Selous (1851–1917), the famous British lion hunter who skinned a considerable number in the course of his career, wrote, 'I cannot see that there is any reason for supposing that more than one species exists . . . The fact is, that between the animal with hardly a vestige of a mane, and the far handsomer but much less common beast with a long flowing black mane, every possible intermediate variety may be found.'[13] He also rejected distinctions based on coat colour, spots on the feet and size, noting that these characteristics also varied considerably from lion to lion, whether female or male.

Scientists today hold that all lions belong to a single species: *Panthera leo*. Various subspecies have been proposed, but a scholarly consensus recognizes eight subspecies distinguished by geographical range and differences in morphology (physical structure):

1. *Panthera leo persica* (the Asian lion)
2. *P.l. senegalensis* (the West and Central African lion)
3. *P.l. nubica* (the East African lion)
4. *P.l. azandica* (the Northeast Congo lion)
5. *P.l. bleyenberghi* (the Southwest African or Katanga lion)
6. *P.l. krugeri* (the Southeast African or Transvaal lion)
7. *P.l. melanochaita* (the Cape Lion)
8. *P.l. leo* (the Barbary or North African lion)

The final two subspecies were hunted to extinction within the last 150 years. Abundant near Cape Town in 1801, there was not a single Cape lion to be found 60 years later. Once ranging from Morocco to Egypt, the Barbary lion, prized by the Romans and featured in bloody entertainments, was extinct in the wild by the mid-twentieth century. Asian lions, now confined to India, were almost wiped out between 1880 and 1920. India's rulers and princes had hunted lions for centuries, but the number of lions did not decrease dramatically until Europeans joined the chase. Showing a typical disregard for their welfare,

The Cape Lion, from Edward Turner Bennett's *The Tower Menagerie* (1829).

a certain Colonel George Acland Smith shot 300 lions in a single year (1857–8). His annual tally, which is possibly exaggerated, approximates the number of Asian lions that survive today.

Recent estimates place the total African lion population between 16,500 and 30,000 animals and they are classified as Vulnerable on the Red List of threatened species compiled by the International Union for Conservation of Nature (IUCN).[14]

The African lion may be more threatened than we think. *P. l. senegalensis*, the lion of west and central Africa, could be classified as 'critically endangered', but because its status as a sub-species has been called into question, its tenuous position has gone unrecognized.[15] How few lions are too few? Evolutionary biologist Pieter Kat argues that 'all predators should be treated as endangered species: they should not have to wait to "deserve" the status when populations have dropped to alarmingly low levels.'[16]

Lions and other carnivores 'make up only 10 per cent of all mammalian genera and only about 2 per cent of all mammalian biomass'.[17] Because lions are poised at the top of the ecological pyramid, they have always been far less plentiful than the animals on which they prey. 1,300,000 wildebeest, 350,000 gazelle, and 200,000 zebra graze in the Masai Mara-Serengeti, for example, but only 3,000 lions pursue them.[18] Like other large, wide-ranging predators, such as the polar bear, the lion is one of the most powerful animals on earth and one of the most vulnerable.

1 Lions at Large

Is it possible for us to comprehend the lion, an animal whose way of life differs so dramatically from our own? Is it true, as Ludwig Wittgenstein suggested, that 'if a lion could talk, we could not understand him'?[1] Our knowledge may never be complete, but this has not deterred wildlife biologists from focusing their binoculars, cameras and microscopes on the great cat, enduring, in their quest for knowledge, extremes of temperature, motor vehicle failures, gastrointestinal parasites, tropical diseases and potentially fatal encounters with their subjects. Several field biologists have almost died of boredom watching lions lying in a languid heap for hours, only to yawn listlessly, roll over and go back to sleep.

Surprisingly little was known of the lion's behaviour and social organization until the late 1960s when George Schaller, the eminent biologist and conservationist, embarked on his research on lions in the Serengeti, Tanzania. He began his field-work in the summer of 1966, and completed it in September 1969, two months after Neil Armstrong set foot on the moon. Schaller's research – a giant leap for lions – was widely disseminated and set the standard in the field. Consequently, many people came to believe that the behaviour of the Serengeti lion was representative of the species as a whole, although Schaller had cautioned against this and anticipated that detailed studies of the animal in other habitats might yield different results.[2]

Since then, lions in Tanzania, Kenya, Uganda, South Africa, Namibia, Botswana, Zimbabwe, Zambia and India have come under scrutiny, but many questions about these extraordinary cats remain unanswered. Lions are idiosyncratic animals that exhibit a remarkably wide range of behaviours. Evidently, they do not read field guides or conform to stereotype. In Botswana, for example, lions have been known to feast on elephants, but prides elsewhere pass on the pachyderm. Animals of both sexes mix in African prides, but males in western India – the sole population left on that subcontinent – rarely consort with females.[3] While most young males of the Tanzanian plains leave their natal prides and strike out on their own, those in Kruger National Park, South Africa, rarely stray far from home.[4] Even within a given habitat, the animals can deviate from one

Because lions have no natural enemies apart from humans, they can sleep undisturbed.

23

Lions have a wide range of facial expressions.

another in actions and appearance. Like the proverbial blind men describing an elephant, field biologists may reach different but equally valid conclusions.

How can we explain variability in lions? The answer lies in the past. Lions are not easily typecast because, over millennia, they have acquired so many different strategies for survival. As we have seen, lions spread from Africa across the globe in a feline diaspora and learned to thrive in various climates and habitats. By the first century, although the lion's range had contracted dramatically, lions could still be found in Africa, Asia and the Near East. Flexibility held the key to survival: populations of lions learned to adapt to prevailing conditions. Consummate killers, they had no natural enemies – people presented the only threat to their existence. But lions are far more than killing machines and to appreciate their complexity it is necessary to consider both their morphology (physical structure) and behaviour.

Unlike some mammals – cows and sheep, for instance – which are more impassive than poker players, lions have a wide range of facial expressions. A lion with its ears erect, eyes closed and drooping lips is a contented cat. A lion with its ears back, eyes narrowed and teeth bared is a furious feline, best observed from a distance. Ear position is of vital importance to inter-lion communication, and the distinctive black markings on the backs of the ears probably help lions detect one another when they are lying camouflaged in the long grass.

Black ear patches help lions distinguish pride mates concealed in the long grass.

Lioness in
Murchison Falls
National Park,
Uganda. Each lion
has a unique set of
whisker spots.

Lioness in Murchison Falls National Park, Uganda. Each lion has a unique set of whisker spots.

Every lion has a unique pattern of dark spots whose number, position and size do not change over the course of its lifetime. These are located above and below the top row of whiskers. In the early 1970s, while conducting research in Nairobi National Park, Judith Rudnai noticed the trait and realized that she could distinguish one lion from another simply by analysing its muzzle. Scars and injuries had previously been used to single out individuals, but these are not foolproof signs. Wounds can heal, two or more lions can experience similar injuries, and cubs may not have suffered wear and tear.[5] The whisker-spot method, a reliable and non-invasive way of identifying individual lions, has been employed by scientists ever since.

Researchers and tourists alike devote large amounts of time, money and effort to view lions in the wild, but animals habituated to humans generally ignore their admirers. As noted by Schaller, 'lions seldom focus for long on another individual. When they do, their eyes have that peculiar detached gaze which gives the human observer the feeling that the lion is looking around and through him rather than at him.'[6] The detached expression is not, of course, due to myopia; lions have excellent eyesight and are also equipped with night vision.

The lion's ability to see in the dark is due to its large pupils and to a light-reflecting layer called the *tapetum lucidum* (bright tapestry), located behind the retina. In photographs taken at night the eyes of a lion may seem to glow, as if illuminated by an internal light source. The *tapetum lucidum*, which bounces

A photograph taken at night shows the effects of the *tapetum lucidum*, which bounces light back through the retina.

light back through the retina, is responsible for this effect, transforming an ordinary animal into the possessed protagonist of a Hollywood horror flick. The oval pupils of the lion contract to a pinpoint, rather than a slit, when subjected to sources of strong light, a feature that distinguishes cats in the Pantherinae subfamily (including lions, tigers, jaguars and leopards), from most of their smaller cousins in the Felinae (a much larger and more diversified group, including lynx, cheetahs and domestic cats).

Cats in the Pantherinae subfamily also have supple larynges (voice boxes), featuring an elastic ligament, the *epihyoideum*, enabling them to roar. Male lions start to roar after their first birthdays and females a few months later, but they do not roar convincingly until the age of two and a half. The sound, a series of plaintive moans building up into a thunderous surge, is one of nature's most compelling, but it is difficult to describe, let alone imitate, and phonetic interpretations leave a lot to be desired. Lions tend to stand while roaring, but they can roar from any position, even at full gallop. Regrettably, domestic house cats are unable to do likewise because their hyoids are inflexible, but they are compensated by an ability to produce a proper purr, a sound that lions cannot achieve. Lions make a purring sound, but it is a half-purr at best, uttered only on exhalation.[7]

Lions roar in response to other lions, to exert their territorial rights and warn intruders to retreat, to enable pride mates to locate them, to intimidate rivals and to defuse tension. If every skirmish between male lions were to escalate into a raging battle the animals could suffer lethal wounds. By roaring instead males are able to signal their dominance and avoid incurring injuries. Communal roaring, behaviour unique to lions, is exhibited by both wild and captive animals, and is believed to strengthen social bonds. Regardless of its precise purpose, roaring can be infectious; other lions, eager to contribute to the tumult, will often join a soloist, forming mixed- or single-sex choirs. Roaring at the break of day, lions form a memorable dawn chorus.

It was recently reported that a lion, incarcerated in a zoo in Baku, Azerbaijan, repeatedly roared, 'Allah'. Supporters of the miracle, who filmed the sequence, saw it as a sign of the creature's devotion to his Creator, but detractors (both Muslim and non-Muslim) pointed out the similarity between the sound emitted by the Arabic-speaking lion of Baku and the standard roar articulated by a lion in the wild.

Stifling a roar can sometimes be more beneficial than expressing it, and lions know when to keep quiet. Studies have shown that nomadic lions, while intruding on the closely patrolled territories of established prides, do not advertise their presence by roaring. In addition, in places where they have been harassed and killed by humans, lions have learned to stay silent to avoid unwanted attention. Furthermore, lions never roar while pursuing their prey.

Sometimes a roar is just a roar.

Pride mates enjoy a communal rest. Panting helps lions regulate their body temperature.

Playback experiments have proved that lions are able to determine the sex of a roaring lion, to distinguish pride mates from strangers, and to recognize individual voices. When females heard recordings of the roars of their male pride mates, for example, they remained calm, but they reacted with alarm to the roars of foreign males. Single lionesses, played tape recordings of three other females roaring in unison, were generally reluctant to approach, but they were much more willing to advance when played a tape of a lone lioness, because they no longer felt outnumbered. These and similar experiments not only shed light on how lions interpret auditory signals, but also suggest that the cats can count.[8] But lions are not mathematical geniuses. Females, for instance, do not use counting skills to keep track of their offspring, and will often abandon a cub or two, inadvertently or otherwise, without acknowledging the loss.

More gregarious than any other great cat, the lion is the only member of Felidae to live in groups and to exhibit marked sexual dimorphism (differences in the form of males and females of the same species). Having inherited territories from their female ancestors, lionesses form the core of the pride, hunting, feeding and raising their young together. Because there is no dominance hierarchy there are no divas or lion queens. A female lion is linked to her birthplace in a fundamental way, and her attachment to her home range never leaves her – she may even prefer to give birth on the very 'spot where she herself was born'.[9] Although lions are more social than any other cats, a pride of lions is not an undifferentiated tawny mass; 'pride members often seem eager to assert their individuality, and the entire pride is very rarely seen together'.[10] If there are too many females in a pride, young ones are likely to be driven away; they may try to stay close to home to establish new prides or they may adopt a peripatetic lifestyle, living alone or teaming up with another lioness.

Pride territories range from 30 to 400 sq km, providing lions with plenty of room to manoeuvre. Males who patrol the home range create a stable social order in which females can raise their young and hunt – a division of labour that enables all members of the pride to flourish. Roles are not firmly fixed. Just as females will help to defend the territory and sometimes kill intruders, male lions make significant contributions by slaughtering big but relatively slow-moving animals, such as buffalo (*Syncerus caffer*), weighing up to four times their own size. Describing a buffalo kill, field biologist Craig Packer writes:

> I once watched a group of females lead their husbands to a buffalo, then stand erect and literally point at the prey: There you are, dears; you can do something useful around the house for once. The males went dutifully forward, hopped on the buffalo's back, rode it like a bucking

bronco, and finally pulled it down. Meanwhile, the females stood perfectly still, cheering the males on from a safe distance.[11]

There is no disputing a male lion's ability to hunt, and it is a skill he must rely on during frequent solitary periods or time spent solely in male company without a female cheerleader in sight. Because mature males weigh over 180 kg (400 lb) (they are 20 to 50 per cent heavier than females), they are also more likely to intimidate scavengers, including hyenas that congregate at kill sites.

Lifelong attachments between pride mates are strengthened by displays of recognition and affection. Male lions generally bond with males, and females with females, rubbing heads, licking, lying back to back, entangling limbs or draping a paw over a neighbour's shoulder. The lion's tongue is a rough rasp; it can strip meat from the bones of its prey, but it is equally adept at dislodging ticks and mites. Mothers and adult daughters like to groom each other, and cubs can receive such an enthusiastic

Grooming strengthens social bonds.

licking that they are knocked off their feet. Hair is ingested during grooming sessions, and lions suffer hairballs just like their smaller cousins. Sometimes the hair will mat and mix with mineral salts forming a hard, polished stone known as a bezoar. Believed by some Africans, such as the Akamba of Kenya, to have magical and protective powers, and to confer on the bearer immunity from lion attacks, the stone was once a treasured talisman. Dying lions were said to spit out the object, and hunters would search diligently for it.

Lions often sniff each other's genitals – a form of greeting that tells them many things, including whether females are ready to mate. Rocks, trees, shrubs and the odd Land Rover are sprayed with urine and the scent from a lion's anal glands, and these markings contain essential data for both scientists and lions alike. As noted by Bruce Patterson, 'anyone who has smelled cat urine knows that it contains plenty of volatile components that have the potential for copious biological information'.[12] Lions make a distinctive scraping motion with their hind feet and often urinate on the ground afterwards. By sniffing pugmarks they are able to track one another's movements and determine whether trespassers have crossed their pride boundaries. When they smell something particularly interesting, like carrion or urine, lions will raise their muzzles, wrinkle their noses, open their mouths and close their eyes – a gesture known as flehmen. This grimace exposes the vomeronasal organ on the roof of the mouth, and enables them to decode scent molecules and pheromones.

Lions' lives are not the frantic and emotionally charged soap operas portrayed in heavily edited wildlife films. In fact, lions spend 80 per cent of their time doing next to nothing. A lion is an overstuffed armchair, a stalled car, a rock. With his eyes half closed he stares into the distance, thinking lion thoughts, dreaming lion dreams. Flies buzz. A single cloud shaped like a

Unconstrained by rules of etiquette, lions never suppress a yawn.

croissant drifts across the sky. There is not an awful lot for a researcher to report. As the indefatigable Schaller monitored the sleepy lions of the Serengeti, Rudnai scrutinized their drowsy counterparts in Nairobi National Park. After measuring intervals between yawns and observing group dynamics, she discovered that a lion can yawn up to five times in twelve minutes, and that yawning is just as contagious in lions as it is in people.[13] Whether the lions had a soporific effect on either scientist remains unclear.

Disappointingly, lions are not inclined to prolonged courtship rituals. There is no leonine equivalent of the leaping and bowing performed by cranes. Biting, cuffing, snarling and growling are standard behaviour for mating lions. What they lack in grace, however, they make up in stamina – they are capable of mating once every twelve minutes for up to six days. Coitus stimulates ovulation; for every successful pregnancy lions may copulate up to 1,500 times.[14] One promiscuous lioness mated so many times that 'her rump sported a smooth, shiny spot from constant wear'.[15] Given the lion's robust sex life, it is not surprising that in some parts of Africa sleeping on a lion's skin is said to increase virility. The German expression for a ladies' man, *Salonlöwen* (living-room lion), likewise alludes to the lion's sexual prowess.

If a female does not conceive initially, she does not have to wait long to try again. Within two or three weeks she will come into oestrus once more and be prepared to present herself to eligible males. As soon as a male has identified a potential mate he will shadow her until she succumbs to his attentions. Strutting males could be expected to fight over mating rights, but they rarely do. Competition for lionesses is governed by the rule of first-come,

Lions have relatively short penises, but this does not discourage them from frequent and vigorous couplings with obliging lionesses.

first-served. The fur may fly, however, if two males reach a female at the same time, or if a lioness, dissatisfied with the efforts of her current partner, sidles up to a male waiting in the wings.

When the time comes for a lioness to give birth she will suddenly slip out of sight. Cubs, born in secrecy and seclusion, enjoy the exclusive company of their mothers for the first six to eight weeks of their lives. Lions are born blind, but within three to fifteen days their blue-grey eyes open, and when they are a few months old these change to amber, complementing their golden pelage. Their fur is marked with spots and rosettes – a legacy of distant ancestors who once lived in woodlands bathed in dappled light. These markings fade as the lion grows older, but vestiges remain on the animal's belly and limbs.

A female calls her cubs with a low grunting sound, but she cannot say, 'Stay here and behave while I catch an impala.' In

The average gestation period for lions is three and a half months. Once they are born, cubs continue to develop quickly. They start to crawl almost immediately and walk within two to three weeks.

order to hunt and mingle with her pride mates, she must slip
away from her cubs undetected. The sense of relief is palpable
when the lioness returns to the den, and the cubs greet her with
chirps, meows and caresses. Seeming to sense their own vulner-
ability, cubs only play freely in the presence of their mothers or
other benevolent adults. Like toddlers, cubs are connoisseurs of
the commonplace. They flirt with danger, splash in rain puddles,
torment tortoises, chase tails. Although growing males abandon childish things, females never lose their sense of fun. Their
frequent interaction with cubs keeps lionesses frisky, and in
times of peace and plenty fully grown females sometimes play
amongst themselves.

When the time is right the lioness will fetch her offspring from
the den and introduce them to other members of the pride.
Lions do not have a breeding season, but females often ovulate

Lionesses in a
pride often ovulate
and conceive at
the same time. A
lioness will suckle
cubs from other lit-
ters in addition to
her own offspring.

37

and conceive at the same time, and raise their cubs communally – the only feline species to do so, and one of the few mammals to engage in this altruistic behaviour. Cubs will suckle at any available teat, and a lactating lioness will acquiesce, but mothers are able to distinguish their own cubs from others, and they are more tolerant of their own offspring. Even so, cubs consume only 70 per cent of their mother's milk and opportunists siphon off the balance. Lion cubs take six to eight months to wean – longer than any other great cat – but they are capable of going without milk for extended periods because lion milk has a high fat content. At about the age of five to eight weeks, a few weeks after lions have acquired their first set of teeth and are ready for a taste of flesh, their mothers will escort them to the kill site. It would seem more efficient to take the food to the cubs, but mothers rarely do, and the exposure to the kill site undoubtedly teaches

A large male makes an ideal climbing frame for a cub in the Masai Mara, Kenya.

the infants lessons we cannot begin to fathom even if we were to stick our heads inside the bloody cavern of a wildebeest's body.

Males do not play a large role in raising their young, but many will patiently endure a drubbing from a pint-sized adversary. Big males make excellent climbing frames, and cubs also take advantage of the shade cast by their father's bulky bodies. Cubs like to imitate their father's actions and will sharpen their claws or yawn at the same time as lion *père*, but their affection and curiosity is often rewarded with a cuff or growl. Irascible males will, however, freely share their food with the youngest cubs, something female pride members refuse to do.

All male lions, and some females, go through nomadic phases in their lives, beginning at the ages of two to four when they leave home. Like human adolescents, some young males go willingly, but others must be forcibly evicted. They are forced to leave by adult pride members, or violently expelled by foreign males during pride takeovers. As a result, these immature and inexperienced lions must forsake the comforting presence of their mothers, aunts and grandmothers. Young females may rejoin their natal prides, but males have no guarantee that they will ever return.

Brothers or cousins often leave the pride simultaneously and stay together as they roam. Since they do not have a territory to protect, or females to guard from rivals, nomadic males may follow migratory herds. They are also free to breed with nomadic females and with unguarded pride females. If they are fortunate enough they may even mate with females whose partners they have overthrown and whose territories they have overtaken – a lion's equivalent of winning the lottery. Evolutionary biologist Pieter Kat observes that 'such dispersing males are critically important in the maintenance of gene flow between increasingly isolated populations of lions'.[16] But male nomads' lives are fraught with danger. For one thing, they invite confrontation and injury when they trespass on territories patrolled by pride males. For another, nomadic lions are likely to wander beyond park boundaries where they risk being killed by hunters or herdsmen protecting their livestock. Confronting these dangers alone is daunting. Some nomads remain independent, but others forge coalitions with unrelated lions – bonds that prove as strong as those between blood relatives. Schaller describes how one solitary lion, which he had tagged and designated No. 57, found a soul mate:

Two young brothers. When it is time to leave the pride, closely related males often stay together.

June 12 [1967] was a momentous day in his life. He had rested all day on a rise and at dusk set out purposefully

on some mysterious errand. At 9:25 p.m. he met a male of
his own age and they walked on together, and by some in-
tangible process they cemented a friendship which lasted
until death.[17]

Unfortunately, death came sooner than expected. On 9 November
1968 a trophy hunter shot the lion when he ventured, alongside
his companion, outside the boundaries of Serengeti Park. 'The
outfitter of the hunt sent the ear tag to me', writes Schaller, 'It
was a generous gesture, yet I cupped the blood-encrusted silver
tag in my palm with a terrible sadness. I would rather have
retained my vision of male No. 57 wandering through his kingdom, the grassy plains, the hollow vastness of the sky. Now I
see him nailed to a wall, staring glassy-eyed, his teeth bared in
supplication.'[18]

Roaring lions sound invincible, but few reach old age and
many die violently, lacerated in fights with rival males, gored by
buffalo or slaughtered by people. Coalitions of nomadic lions
seek to find and overthrow adult males associated with a pride,
causing the latter to be ever vigilant, patrolling the boundaries of
their adopted home ranges and cutting off access to their female
pride mates. There are exceptions to the rule, but even the most
robust lion does not usually dominate a pride for more than two
or three years before being forced into exile by younger or more
powerful males. Drawing on his research in Tanzania, Packer
states, 'for lions, the conflict never ends: the configuration of territories that we see today reflects an endless past fractioning of
clans, the balkanization of the Serengeti'.[19] Wherever prides of
lions exist, territorial boundaries shift in accordance with pride
membership and the balance of power between male overlords.

Like male cougars, leopards and tigers, lions also kill, and
sometimes eat, their rivals' offspring. Their goal is not to obtain

Lions fighting in the Okavango, Botswana. Conflicts between lions may result in lethal wounds. Healthy males weigh an average of 188 kg (415 pounds).

a protein supplement but to create reproductive opportunities for themselves. Cubs suffer high mortality rates: at least 50 per cent die before reaching maturity.[20] 'Infanticide is a vital part of male reproductive strategy' – it is how lions are programmed to perform.[21] Although they have been known to come into oestrus when their cubs are still suckling, females do not usually mate and have another litter until their cubs are at least a year old.[22] Nomadic males, with no time to spare in the race to disseminate their genes, cannot afford to wait for the females to become receptive to mating, so they generally kill the infants they encounter when they seize control of a pride.

Ongoing research has shed new light on infanticide in lions, including its correlation with female sociability. It has been suggested that lionesses in general, and mothers in particular, are highly social so they can best defend their cubs against murderous males. Single mothers would be vulnerable in isolation, but by banding together they improve their chances of protecting their young. Experiments have shown that female lions roaring in unison are able to intimidate potential male interlopers passing through their territory and to minimize the risks of attacks on their offspring.[23]

Lions will eat whatever prey is present in abundance and in many places their diets fluctuate with the seasonal migrations of herds. In Kruger National Park, South Africa, lions can choose from 37 different types of prey and in the Okavango Delta, Botswana, they consume at least 19 different kinds, 'ranging in size from springhare to giraffe and elephant'.[24] In the Serengeti lions prefer three main staples – zebra, wildebeest and Thomson gazelle – while their Asiatic cousins in India feast on wild boar and deer, among other animals. The carnivores generally scorn fruit and vegetables, but lions in the Kalahari Desert quench their thirst with *tsamma* melons and gemsbok cucumbers.

Two females enjoy a feast in the Okavango, Botswana. Lions consume prodigious quantities of meat.

Photographs taken by Beverly Joubert in Chobe National Park, Botswana, show lions feasting on elephants at a waterhole.[25] Evidently, the lions of Chobe taught themselves how to send the multi-ton animals crashing to the ground. Data collected between 1993 and 1996 indicates that the lions killed 74 elephants, with the numbers rising annually as the cats refined their techniques, learning, for example, how to stay out of trunk's reach.

If a single lion has learned a skill, she can pass it on to pride mates. In the Okavango Delta, Botswana, 'the Mogogelo pride have learned that treed baboons rarely remain in their safe perches: when a lion makes a rush up the base of the tree they panic and leap to the ground, where they are easily caught'.[26] The lions do not need to climb the tree, feigning a lunge is enough to bring the baboons down. Other nearby prides do not engage in this behaviour, so it is clear that hunting baboons is an acquired skill. A pride of lions living on the Skeleton Coast in Namibia in the mid-1980s learned to hunt seal, but the art died out when the last of these lions was destroyed.[27]

Lions engage in ferocious bouts of gnashing and growling, clawing and biting at every meal. The sight and sound of lions threatening to tear each other apart can be disconcerting, although little lasting damage is usually sustained. Once they have elbowed aside the competition, lions shear off the flesh with their teeth, bolting down great lumps without chewing, and staining their faces and chests with blood. Male lions do indeed gain 'the lion's share' whenever they can muscle in and snatch away the kill, and since there is no defined hierarchical ranking each lion is left to fend for him or herself: 'it is a system based on the amount of damage each animal can inflict on an aggressor'.[28]

Even when there is an abundance of food lions act this way, which suggests that at an earlier point in their evolutionary history fighting for food was crucial for survival. No concessions are made for cubs and young lions, and these are the first to go hungry when prey is scarce. Until she has satisfied her own appetite a mother will not share with her offspring, even though they may starve to death. By contrast an African hunting dog will regurgitate meat for her pups and ensure that older offspring have the first chance to eat at a kill.

The distended belly of a satiated lion is an impressive sight, lending additional breadth to an already bulky beast. Sometimes a lion that has overeaten will stagger to its feet and make another kill. This behaviour suggests that the urge to eat and the urge to stalk exist as independent drives in lions, as they do in domestic cats, which are notorious for preying on songbirds even when well fed.

Unlike house cats, however, which cover their faeces, lions make no attempts to conceal theirs. Scientists can learn a lot from lion dung, but not before donning a pair of gloves. Liver flukes, hookworm and giardia are some of the intestinal parasites found in lion faeces, alongside toxoplasmosis and echinococcus – a tapeworm that can cause cysts up to twelve inches in diameter.[29] Lions that have consumed a lot of blood will produce dark, foul-smelling faeces, and hair and other indigestible matter, including porcupine quills, may be embedded in the waste. Incredibly, there is a healthy market for lion faeces. Pellets soaked in the essence of lion dung are said by satisfied customers to deter domestic cats from gardens. According to advertisements 'even the bravest cats will retreat when they smell a lion'. The lions' meat-based diet not only affects the pungency of their waste, but also contributes to their notoriously bad breath, a characteristic noted by Pliny the Elder (23–79 CE) and countless lion-tamers.

Ever opportunistic, hunting lions will use safari vehicles as cover, wait in ambush near sources of water and attack prey startled by storms. Once, when Lord Delamere (1870–1931) was galloping after a wounded oryx, a lion sprang up to snatch his prize away, pulling it down under his very nose.[30] Lions will never refuse a free meal and respond with alacrity to circling vultures and cackling hyenas.

The aesthetically challenged hyena is much maligned. It was long believed that when both lions and hyenas were present at a kill the latter invariably stole from the former, but this has proved to be untrue; we now know that the lion is just as inclined to pilfer the hard-won prey of the hyena. When hyenas and lions compete for resources the balance of power between them depends on a number of factors, including relative size of clan versus pride and the availability of food. In a study of lions in Etosha National Park, Namibia, conducted in 2000–1, spotted hyenas (*Crocuta crocuta*) never succeeded in wresting kills from lions, and always failed to protect their own food. However, hyenas in the Serengeti and elsewhere, which were the subjects of earlier research, were not persecuted to the same extent by the domineering cats. In fact, hyenas in the Ngorongoro Crater,

Hyenas and vultures attract lions to kills and vice versa.

Tanzania, were able to appropriate the lions' kills 100 per cent of the time when no male lions were present.[31]

When the hunt is successful, it is as if a marvellous internal clock cues the cat to spring at precisely the right moment; whether we identify with the prey or the predator, it is a mesmerizing spectacle. Our fascination with the kill is rooted in our evolutionary history, in our latent admiration for animals skilled in hunting, an activity that we too have had to master in order to survive. Our earliest ancestors, once they had learned to hunt, had as their closest neighbours 'well-established, well-adapted, and well-armed predators, all doing the same thing in the same habitat'.[32] Early hominids may not have borrowed garden shears or cups of sugar from the lions living next door, but they may have borrowed a few hunting strategies, and they scavenged meat from fresh kills.

Despite punishing temperatures, lions can survive for relatively long periods without water. Because they have low metabolic rates, they can also endure prolonged fasts. The animals rarely feed while mating, for instance, but will gorge themselves to bursting at every other opportunity. Lions have better timing than most stand-up comedians, but hunts often end in failure. Dodging flying hooves and horns is a dangerous business and lions often suffer injuries or losses to their eyes, ears and teeth – injuries that may lead to starvation or septicaemia. Worn or missing teeth are especially common in geriatric lions, but animals of any age are susceptible to abscesses and dental traumas.

There is no scholarly consensus regarding the purpose of group hunting in lions – a fundamental issue since no other cats engage in this behaviour. Field biologists have noted that lions, like members of a football team, sometimes assume distinct roles, which they play repeatedly.[33] In some instances, cooperating is an efficient strategy, but group hunting does not necessarily

It takes several lionesses to bring down an African buffalo (*Syncerus caffer*), a dangerous, and unpredictable opponent.

confer advantages. Two lions hunting together stand a greater chance of success than an independent hunter, but adding additional lions to the hunting party does not improve the odds. Often it is a case of too many cooks spoiling the broth. Lions can miscalculate distances or dash forward at inopportune moments, seemingly oblivious to their colleagues' efforts. Even when a group hunt is successful, a deceitful lion may pretend to fail, concealing her kill in the long grass until her disappointed pride mates have dispersed.[34] Lions hunt collectively, but not always cooperatively. If hunting en masse is inefficient why do they do it?

Some scientists have suggested that because lions are the only great cats to hunt in groups, and are the most sociable felids, these two aspects of their behaviour are linked: lions stay together to hunt together. Others have posited that female lions – the core of the pride – do not form groups in order to hunt collectively, but to fend off male interlopers, thereby protecting their cubs. Still others believe that group hunting is actually effective. 'It may well be', writes carnivore expert Hans Kruuk, 'that in a cooperative hunt the participants have to work less hard than in a solitary effort . . . and in that case collaboration would pay.'[35] Females hunt more often than males but, contrary to popular belief, this is not because the latter are lazy chauvinists, but because the lighter, more agile females have a greater chance of success. Males are not only heavier and less nimble, but their manes, which are as conspicuous as 'moving haystacks', give them away.[36]

The mane is the lion's most distinctive feature. This hairy appurtenance, along with the lion's tail tuft and uniform pelage, sets the animal apart from all other felids. Made up of guard hairs (coarse, water-shedding hairs), manes start to emerge when males are about a year old and continue to grow as the lion ages. Testosterone stimulates mane development and castration inhibits growth. Genetics influence mane colour and thickness, but the appearance of a lion's mane can change over the course of his lifetime: a sickly or injured male may shed mane hair, whereas a mature, healthy lion living in a rich habitat may sport a luxuriant mane. Needless to say, females do not have manes, but 'occasionally, elderly females develop shaggy sideburns'.[37]

Asiatic lions have less luxuriant manes than their African counterparts and the hair is sparser on the tops of their heads, which makes their ears seem more prominent. Not all Asiatic males look the same, however. The Inuit do not have as many words for snow as once rumoured, but the Maldharis, pastoralists living in the Gir forest in western India, have several different words for male lions, depending on the colour of their manes: 'a full grown lion with a light brown mane is called *bhurio*, a lion with a light coloured (yellowish) mane a *pilo*, and a black maned lion a *kalio* or *kamho*'. The term *radiyo*, sometimes applied to a noisy person, is also used to designate a lion prone to roaring.[38]

Although scientists agree that the lion's mane serves an important function, they are not certain of its precise purpose. It was long believed that manes protected the necks of fighting males, but this theory has been challenged because battling lions do not target the nape. A series of experiments conducted in 2002 by Peyton West and Craig Packer suggest that the mane advertises the lion's physical and genetic fitness. Four life-size plush lions were introduced to female lions in the Serengeti. Two sported blond manes of varying lengths, and two dark. The lionesses showed a distinct preference for the dark-maned mannequins, and ignored the blond surrogates. The length of the mane was of little significance to the females in the study group. Males with dark manes do have higher testosterone

A lion with a luxuriant mane in Serengeti National Park, Tanzania.

A lion's tail can measure up to 105 cm (41 inches) long. The precise purpose of the tuft has eluded scientists.

levels, dominate other cats at kills and consume more calories. Hence, the females' attraction to the dark-maned dummies has an underlying logic: 'By preferring males with darker manes, females gain mature, better-fed, more aggressive mates.'[39] In short, the mane appears to be a sexually selected trait and, when it comes to competition for females, lions with dark manes have an advantage.

But climatic conditions also exert a major influence on mane growth. Free-ranging males in relatively cool regions and males in captivity in the northern hemisphere often have long, luxuriant manes. Conversely, lions living in hot, dry regions, where having a thick, dark mane would be a liability, have sparser ones. Lions in Tsavo East National Park, Kenya, for example, have feeble facial hair rather than substantial manes. East Tsavo lions are, however, able to attract mates, so it would seem that for this population at least, mane characteristics are not reliable indicators of fitness or reproductive potential. The arrested mane development of the East Tsavo lions reflects their dry habitat: 'no rain, no mane'.[40] An East Tsavo lion will sprout a full mane if transplanted to a less arid environment.

While research on the lion's mane advances, the enigmatic tail tuft awaits its champion. The black furry tuft is absent at birth, but begins to grow when the lion is about five and a half months old. Found in no other felid, the tuft conceals a short section of fused bone at the tail's tip, a horny spur, which is equally mysterious.

Lions are as individualistic as we are: some are irritable and aggressive, others friendly and peaceable. Although they generally avoid anthropomorphic interpretations of animals, seasoned field biologists, seduced by the individual quirks of the big cats, often give them evocative or silly names. This has the unintended effect of making field notes – qualitative data based on hours of painstaking observation – read like pulp fiction: 'Blondie – it must be admitted – proved to be a shockingly irresponsible mother . . . she had cubs in July, but in August was already seen flirting with one of the males.'[41] 'When Patricia, Romola and Chryse killed a warthog, Patricia, who was keeping a stranglehold on the animal's neck . . . did not start feeding immediately.'[42] The expendable warthog is simply a snack, but Patricia, Romola and Chryse are individuals, able to think and act for themselves.

Studies of prides are based on thousands of hours in the field, so it is not surprising that wildlife biologists form a bond with 'their' lions. Lions can get under the skin of the most objective scientist; the more they grow to recognize and know individual lions the more emotionally engaged they become. 'One cannot help empathising with their problems,' notes one researcher, 'mourn the loss of cubs, keep a concerned watch on animals that have been wounded, and laugh at the comedy of playing cubs and bumbled hunts.'[43] How can we maintain a critical distance? It helps to remember that our fascination with every aspect of the lion's life is not reciprocated. Lions invariably retreat when humans approach on foot and, judging from their indifferent mien, they are quite content without us.

2 Captive Cats

It is one of the most famous wildlife stories of the twentieth century. George and Joy Adamson, living in northern Kenya, adopt three orphaned lion cubs. Two are sent to Rotterdam Zoo but the runt of the litter, Elsa, lives with them until she reaches maturity and they return her to the wild. Elsa learns to hunt, mates with a wild lion, gives birth to three cubs, and then dies suddenly. It is a straightforward account of the relationship between a game warden, his wife and a lioness. Or is it?

Elsa's story is full of contradictions. It was George Adamson's job to shoot 'troublesome' lions that preyed on local people or livestock. Elsa's mother, a victim of mistaken identity, had been killed on just such an expedition when she had sprung out of her lair to protect her cubs. After destroying the lioness, George preserved her offspring; within days of her birth, Elsa was sucking milk from a bottle and being coddled by the Adamsons. As suggested by the title of Joy's best-selling book, *Born Free: A Lioness of Two Worlds*, Elsa occupied an equivocal position. Fully habituated to humans, she nevertheless retained her natural instincts. Not entirely at home in either world, she drifted between the two.

Joy and George insisted that the lioness was not a pet, but hundreds of photographs undermine these claims. Here, Elsa squirms in Joy's arms; there, she fishes with George and carries home the catch. Upside down with her paws in the air, Elsa dozes on a camp bed, and draped over the roof of a Land Rover, she catches the evening breeze. These sensational photographs which Joy, oblivious to the irony, kept in an album bound in lion skin, guaranteed the success of her book and made Elsa famous. Within months of its publication in 1960, *Born Free* was translated into 25 languages and it earned half a million pounds in its first ten years in print.[1] Joy immediately established a charitable trust to help protect wild animals, and suddenly people who had never given lions a second thought joined the conservation movement.

The universal appeal of *Born Free*, its sequels and the Hollywood film of the same name rested on the domestic idyll projected in the photographs; here was a couple, without any specialized training with animals, living with a lioness who seemed as docile as a cat. Whenever Elsa was nervous, she would suck Joy's thumbs, and she followed George around like a dog. Joy would even administer enemas when Elsa was constipated, an

Joy Adamson with Elsa.

indignity to which few house cats would submit. But the harmonious relationship presented in *Born Free* did not always reflect real life; living with a wild lion was more difficult than it looked. Once in a while Elsa acted aggressively, but these moments are never alluded to in the books or the film, which present, instead, an idealized picture of an affectionate animal.

Adopting Elsa forced the Adamsons to renegotiate their relationship with lions, but George continued to shoot them when required, and Joy and Elsa would sometimes accompany him on safaris in pursuit of wild lions that had harassed local herdsmen. On one occasion, after a lion had been shot and skinned, Joy photographed its heart. She writes, 'it was as big as a child's head; then I understood why I often felt Elsa's heart throbbing like a motor engine against her ribs'.[2] Joy associated the dismembered wild lion with her pet lioness, but she was able to reconcile her conflicting attitudes towards the two animals. *Born Free* chronicles Elsa's life, but Joy, who comes across as an ardent yet unreflective animal lover, devotes little space to the bigger questions of inter-species relations, or to the place of human beings in the natural order.

Having spent the first half of Elsa's life trying to suppress her natural instincts and allowing her to rely on them for food and company, Joy and George spent the last half encouraging her to redeploy her latent urges, to learn to hunt, and to achieve independence. In the three years in which they lived with Elsa the Adamsons grew to understand her character, and George continued to work with lions for the rest of his life, gaining an unparalleled knowledge of their behaviour. But when they released Elsa into the wild in 1959 little was known about the social organization of lions, and the Adamsons failed to grasp how difficult it would be to persuade wild lions to confer pride membership on a tame stranger. It is not surprising that Elsa was at first baffled by their attempts to introduce her to wild lions. Releasing a tame lion also presented other problems. Because the animal had lost its fear of people, it was liable to approach settlements, stalk livestock and either kill or be killed.

During the making of the movie *Born Free* (Columbia Pictures, 1966), which required 21 different leonine actors, the human participants, including the stars Virginia McKenna and Bill Travers, learned how intractable lions could be. Even with George Adamson on set to coach the cats, the simplest scene could prove demanding. 'You just want the lion to sit there between Bill and Virginia for a minute and you could be a week on that', despaired director James Hill.[3] And working with lions had other drawbacks. George's favourite animal, a lion named Boy who gave a commanding performance, injured McKenna on the set and later, in Adamson's custody, mauled a child. In 1971 Adamson finally had to shoot Boy when he killed one of his African assistants, a man who had doted on the animal for many years.

Virginia McKenna and Bill Travers in *Born Free* (1966).

When he was introduced to George Adamson in the late 1960s the biologist George Schaller, who had spent countless hours observing wild lions, acknowledged that by raising a cub, Adamson had entered fully 'into the lion's world', and 'knew lions as individuals better than anyone' he had ever met.[4] Proximity was crucial; entering the lion's world meant getting as close to them as possible.

For lion-tamers, this was a matter of course: they climbed into the cage with lions and learned to anticipate their every move. One of the very first to perform with the felines was the American animal trainer Isaac Van Amburgh (d. 1865), who entertained crowds on both sides of the Atlantic. Dressed like a Roman gladiator, Van Amburgh would subdue by force a mixed group of big cats and then lie in their midst. 'One can never see it too often', wrote Queen Victoria in her journal, and she proved her point by going to see him perform six times in just over a month when he was in London in 1839.[5] She also went backstage to see the animals being fed, and commissioned Edwin Landseer to paint a portrait of the 'brute tamer' in action. Van Amburgh, who beat his charges into submission with a crowbar, lacked

Lions play with their keeper, Circus Rebernigg.

subtlety but he did not lack confidence. He would bring a lamb or a child into the cage to emphasize his authority over the animals and to allude to the biblical prophecy of a Peaceable Kingdom, and his act included 'forcing the lions to approach and lick his boots as the ultimate sign of his conquest and the animals' abject subservience'.[6] It is not surprising that Van Amburgh's performance struck a chord with Queen Victoria. While Van Amburgh was taming his lions and tigers, the British lion, a wide-ranging imperial beast, was poised to conquer the world; the 'scriptural sanction for dominion' was being superseded 'by the secular ideology of imperial conquest'.[7]

Who was the first performer to stick his head inside a lion's mouth? It is difficult to say with certainty, but Van Amburgh is a strong contender. Surprisingly, a jaded public soon regarded the trick as routine. Writing in 1882, a journalist stated, 'When

Edwin Landseer, *Isaac Van Amburgh and his Animals*, 1839, oil on canvas; commissioned by Queen Victoria.

Van Amburgh used to put his head in the lion's mouth in the earlier days it was considered a wonderful and foolhardy thing, but people have become accustomed to the sight and do not marvel as much as they did.'[8] Another, reporting in 1909, refers to the trick as 'ancient and hoary', and reminds his readers that a lion tamer who attempts it must 'keep his hair trimmed short to avoid irritating the inside of the lion's mouth'.[9]

White Americans and Europeans dominated the profession in the early days, but some black lion-tamers achieved celebrity. The West Indian John (or Alexander) Humphreys, born on the island of St Vincent in 1859, travelled to Britain as a ship's boy, and made his debut there as a lion-tamer before he turned twenty. Humphreys, known by the stage name Alicamousa, performed with several lions, including the 2.4m (8 ft) long Wallace whom he pretended to overpower. Despite Wallace's impressive physique, the act was convincing since Alicamousa, an equally imposing figure, stood over 1.8m (6 ft) tall in his stocking feet.

During a show in Birmingham in 1881 Wallace lashed out at his trainer. Although Alicamousa was not seriously injured, the sudden attack by the lion, which lacerated his scalp and right arm, panicked the crowd. After having his wounds dressed, the stalwart Alicamousa wished to go on with the show, but he was overruled, and the unreliable Wallace was relegated to retirement.[10] Press reports state that 3,000 people were in attendance and two-thirds were children. The fact that such large numbers of youngsters were present suggests that lion-taming acts, despite their threat of violence and potential for cruelty, were deemed appropriate for children. Only in recent times has this premise been questioned.

By the second half of the nineteenth century two main styles of wild animal performances had emerged: *en férocité* and *en douceur*. Trainers conforming to the first style, like Van Amburgh, adopted a combative approach, dominating the animals through a series of aggressive postures, whereas trainers conforming to the second stressed their familiarity with the great cats and used quieter commands to demonstrate their mastery over the animals.

The most influential animal trainer of the period was Carl Hagenbeck (d. 1913), based in Hamburg, Germany, who is credited with two key developments: training animals without using excessive force and allowing them to perform in larger arenas with a variety of props instead of in cramped cages. Lion training was a Hagenbeck speciality. In 1893, after making a name for himself in Europe, he took his big cats to Chicago where they, and a large cast of other animals, entertained at least a million people at the Chicago World's Fair. Hagenbeck's lions did spectacular things no lion had ever conceived of doing before: rode horses, rode tricycles, played on a seesaw, and served as hurdles for dogs. They also took their place in 'The Great Zoological Pyramid' alongside tigers, panthers, leopards, bears and dogs. The grand finale consisted of a crowned lion 'driving' a chariot drawn by two tigers, with a pair of Great Danes on the running

Carl Hagenbeck (1844–1913), German zookeeper, circus impresario, and animal dealer.

The triumphal procession of the Lion-Prince, Adolph Friedländer poster for Hagenbeck's Circus, 1895–6.

Heinrich Mehrmann performing with Hagenbeck's Circus, Friedländer poster, 1896.

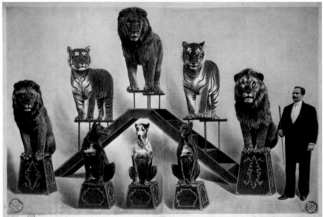

board serving as footmen. With so many different animals working in concert, the Hagenbeck Pavilion seemed to fulfil the promise of the Peaceable Kingdom. Hagenbeck's trainers, including his brother-in-law, Heinrich Mehrmann, reinforced this impression. As noted by the scholar Nigel Rothfels,

Hagenbeck's performances represent a very new way of imagining animal–animal and animal–human interaction. Instead of a gladiator beating the beasts into submission, here we see a former Hamburg businessman, Heinrich Mehrmann, as he stands respectfully in evening wear beside his calm, far-from-roaring charges.[11]

Professor Deyerling, another animal trainer employed by Hagenbeck, would hitch three lions to his chariot, and, as a finale, recline on top of his group of lions, as if he were stretched out on a couch.

Several female lion-tamers became famous for their ability to control their animals with gentle words and subtle gestures.

Claire Heliot shares a tranquil moment with one of her charges, Friedländer poster, 1903.

Claire Heliot's demure appearance belied her strength; postcard, c. 1902.

Dressed in a white satin evening gown, the German lion-tamer Clara Pleßke (1866–1953), who toured Europe, Russia and the United States under the stage name Claire Heliot, used charm rather than brute force to subdue her animals. 'When they throw their ears back and snarl and reach out for her in lightning swiftness with one paw she coquettes with them', wrote a New York Times journalist in 1905. 'Coquetry must be a remote possibility to a lion, but Claire Heliot relies upon his appreciation of it, nevertheless.'[12]

Eager to appeal to a conservative public with fixed ideas about suitable roles for women, journalists and publicists alike emphasized the feminine qualities of Heliot's performances and that of other early women lion-tamers. Although Heliot was strong enough to carry on her back a 159 kg (350 lb) lion, her delicacy and maternal instincts were stressed rather than her strength; the 'little woman' was said to feed and bathe her fourteen 'pets' every night and to govern the cats with pats and caresses rather than brute force. Dinner parties, presided over by Claire, who sat at table with her well-mannered lions and offered them choice morsels, enhanced the image of domestic harmony.

A woman in a cage with fourteen lions was an irresistible prospect; people wanted to see for themselves if Claire would receive a playful pawing from her favourite Sascha or a nip from the ill-tempered Auguste. 'Of Which One of Her Fourteen Pets Will She Be the Victim?' ran the headlines, and people flocked to her performances to find out.

Another female lion-tamer who found fame at the turn of the century was Mademoiselle Adgie, who travelled across America with her feline troupe, including the lioness Victoria, a 'sulky cat, with a skulk like an assassin and teeth like whetted knives'.[13] Unlike the demure Claire Heliot, Adgie would sometimes strip to her undergarments and was celebrated for dancing the tango with a lion.

In Chicago, in June of 1914, tragedy struck. A large male lion named Teddy tore to pieces Adgie's assistant and lover, Emerson D. Dietrich. Since the 24-year-old Dietrich died in suspicious circumstances, Adgie was implicated in his death, but a coroner's jury deemed it an accident. Focusing on Adgie's relationship with her lions, reporters claimed that Teddy had killed Dietrich

Mademoiselle Adgie poses with her troupe, c. 1899.

in a fit of jealousy, and Adgie encouraged this view, stating: 'Lions are just like dogs or cats. They know when some one else shares the affection of their owner. Now that I think of it, it was only a few weeks ago that Emerson put his hand caressingly on my shoulder on the stage at Salt Lake City. I looked over at Teddy . . . His eyes glittered, and he snarled.'[14]

A few months after the accident, in August of 1914, an article advertising Adgie's show at the Palace Theatre, New York, reflected Teddy's new billing: 'Adgie's lions, including the man-eating Teddy, will be the new headliners. There are ten lions in the act, which is the most thrilling animal exhibition ever given on the stage.'[15] An erotic undercurrent gave an additional charge to Adgie's performances, but this was relatively understated. In subsequent decades, however, whip-wielding female lion-tamers dressed in tight white breeches, low-cut blouses and knee-high boots were cast in the role of glamorous dominatrices.

Some late nineteenth-century lion-taming acts combined the ferocious and the docile. Visitors to Coney Island, New York, including the young Mae West, thrilled to see Captain Bonavita, dressed in a military uniform, issuing imperious commands to his leonine subordinates. Although he would crack the whip, his act included a routine called 'The Old Arm-Chair' in which

'The Old Arm-Chair', from Frank C. Bostock's *The Training of Wild Animals* (1903).

he transformed his sizable troupe into a tawny pyramid and sat calmly in their midst reading a newspaper. Another routine, similar to Heliot's, consisted of the Captain carrying a 225 kg (500 lb) lion on his shoulders. In a photograph showing the lion balanced on his back, Bonavita, sporting an impressive handlebar moustache, maintains his military bearing despite the beast's tremendous weight. The photograph documents the man's control over the animal, but it is the massive and inscrutable lion, staring impassively at the camera, that holds the viewer's gaze. Proximity to lions came at a price. In 1904, after being clawed by a lion named Baltimore, Captain Bonavita had his infected arm amputated. Although he returned to the ring and drew even bigger crowds as a one-armed lion-tamer, he eventually died in 1917 of injuries inflicted by a polar bear.

Memoirs by lion-tamers who lived long enough to write them stress both the animal's unpredictability and individuality. In his book *The Training of Wild Animals*, published in 1903,

'Captain Bonavita Carrying a Lion Weighing Five Hundred Pounds', from Frank C. Bostock's *The Training of Wild Animals* (1903).

Frank Bostock describes a lion that was intimidated by a stick held in the left hand, but scorned any number of weapons held in the right. 'No satisfactory explanation of this individual peculiarity has ever been offered,' writes Bostock, 'and one trainer limps for life simply because he did not make the discovery in time.'[16] One lion objected to a certain piece of music and was so disruptive that the orchestra had to choose an alternate tune. Another disliked his trainer's gaudy new costume and tore it off his back. Sensitivity to sartorial excess is a recurring theme. The celebrated British trainer, Patricia Bourne, noted that her lions had 'a certain dress-sense' and recalled how worried she had been when an elaborately attired female journalist insisted on entering their cage.[17] A lion's mood could also fluctuate with the barometer. Damp, muggy weather made for irritable lions, and hot days for listless ones. Plummeting temperatures could, however, serve as a tonic.

A circus lion had to earn its keep: its value was contingent on mastering a series of tricks and performing these consistently. Lions like people, had various aptitudes and abilities. Some took a week to learn a new trick that another could master in a day. In his memoir *No Bars Between*, published in 1957, Alex Kerr, a Scottish lion-tamer with Bertram Mills Circus, states that some of his lions had trouble remembering their names. Explaining how he would refresh their memories, he writes: 'If it looks as if one of them is forgetting his name I take him alone in the cage, seat him on his prop, and run through a fake series, among which is his own, allowing him to drop down only when it is called. Atlas, for instance was always trying to belt away when I called 'Suleiman'. I told him firmly and loudly, 'Your name is *At*-las!'[18] Perhaps Kerr's Glaswegian accent was responsible for some of the confusion.

Trainers state repeatedly that no two lions are alike. Rajah, for example, one of Kerr's top stars, 'loathed anybody to be jolly and friendly in the mornings, until he felt sufficiently awake to face the world'.[19] And only a single lion, a male named Belmonte, tolerated being ridden by Patricia Bourne. Belmonte's skin, like that of all lions, was very loose, and it tended to sag and shift, making it hard to stay on top of the big cat. Once, Bourne did slide off his back and Belmonte seized her arm in his jaws, but he bashfully 'gazed at his toes' when she scolded him like a cross schoolteacher, 'Monty! Really, whatever are you doing?'[20]

Can animals perform in the same way as human actors? Do lions know that they are on display? Since animals, including lions, 'are normally held to be incapable of conscious deception' it would seem unlikely that they can perform in a conventional sense.[21] But lion trainers have noted that many animals respond positively to the audience's adulation and seem to enjoy learning their routines. According to Kerr, Rajah, a consummate performer and crowd pleaser, sank into depression when, overcome by ill health and old age, he was forced to sit on the sidelines, listening to the audience applaud younger, more supple lions.

Unlike many lion-tamers, Kerr advocated a 'quiet approach' and did not goad his animals or fight with them in the ring. His methods were so effective that he even trained a sizeable male called Negus to walk a tightrope suspended 1.8m (6 ft) off the ground. Another lion had been slated to do the trick but Negus, intrigued by the challenge, usurped the role. However, many other lion-tamers, notably the American Clyde Beatty (1903–1965), played up the lion's reputation for savagery. Beatty would appear to incite mixed groups of lions and tigers so that, armed with a whip and a pistol loaded with blanks, he could do mock battle with up to 40 of them at one time. The controlled violence of the fighting act appealed to the public, and the image of the lion-tamer staving off a lion with a dining chair owes itself to the influential Beatty, who went on to achieve

Negus walks the tightrope at Bertram Mills Circus, Olympia, c. 1950.

celebrity as an actor in B-movies and television serials. Ernest Hemingway was so impressed by Beatty's mastery of the big cats that he gave him a signed copy of his paean to bullfighting, *Death in the Afternoon*. Although the fighting act was nothing new, Clyde Beatty's lion-taming performances were more spectacular than those of his predecessors and he set the standard for several decades.

As a young man Dave Hoover was so fascinated by Beatty's act that, against his mother's wishes but with the sanction of his priest, he joined Beatty's circus and took up lion-taming. In Errol Morris' documentary *Fast, Cheap and Out of Control* (Sony Pictures Classics, 1997), Hoover sheds light on his days in the ring. According to Hoover, when a lion is confronted with a chair the four legs present a complex spatial problem. He states: 'an animal has a one-track mind. For instance, the animal is coming after you with the idea of tearing your head off . . . You put the chair up, and, all of a sudden, he has four points of interest . . . His mind now has been completely distracted from his original thought: Eat the man in the white pants.'[22] Are performing lions as dedicated to destruction as Hoover suggests? It is difficult to say with certainty. No matter how much we desire to enter the lion's world, we can only go so far. But Hoover's assertion that a trainer is incapable of stopping even one attacking lion is undeniably true, and his parting advice is worth remembering: 'Never let that bluff down.'

A shoemaker's son, who has had too much to drink, rides a lion that has escaped from the king's menagerie, from a copy of the *Shahnama*, 1616.

The wild-animal act is more complicated than it may at first appear. On the one hand, potentially lethal predators are trained to sublimate their aggressive tendencies and antipathy to humans, but on the other they are encouraged to snarl on cue. A lion in a fighting act cannot *be* antagonistic, but it must *act* antagonistic. The animal is complicit in the theatrical performance but, unlike the lion-tamer, it is not conscious of its participation. In

یا رسی رین زوده اندرسو
پیاده هر خانه سوراخ کرد
یکی شیر خست برآ
پیاده و بگرفت کشش
یک دست زبیرو دکشکند
دلیراندر آمد نبرد یک شیر
زده مه و نخرد ازار نخواند

کمر شکنی بابانی آمد
جوانمرد در جام خسته گر
خان بسی کزخانه شیران شاد
بشد تیر میرو برشیر خست
سمی شد نوان شیران خون
پیاده دمان تا دربار کا
جهانداران از انکه شکنی ماند

حورزا شوی المن بشاد کام
سهابی پوپیش سخن خوش
شد شاد وابا فده را خوش
پیاده اکشتا وشتم سود
غلام ابرو شیر زربو
نشستہ خور خرسوار دلیر
بدید بدید آنک شنید بود

سور جوان کفت ابنین شام
بزدکشکریان سی منت
وزابجاک هشدد مدکاه خوش
انان سی سمی کشک جست سود
ماکه شیر بله سیه سود
یکی کشک کرده برشتیم
بمنت آن شکستی کیاد بود

بوید جین کفت کین کشکر
بجستند وکشند بابا درش
همان درشکون تحن رخ درآ

نحکین کن که تابازه کدازه کهر
فایده که برمنة کوشش
بپیش شهشاه برکشت راز

اکر پهلوان زراده بانشدروا
نیاکنت کرد بدرکشکر
که ای نبار کرد نکل لشکر

کبر پهلوان دلیری سرا
دران پشه برنیا ید کهر
توشادان سری با یو دروز کا

reality, the lion-tamer is no more powerful than the Wizard of Oz. If the tamer permits the lion to peek behind the curtain (or chair), the show is over.

Modern zoos have their origins in royal menageries, established by rulers to convey their dominion over all living things. The symbolic association of lions with kingship, dating back millennia, made them perfect diplomatic gifts and royal mascots. As early as the eighth century BCE, Assyrian kings bred captive lions and they also engaged in ritualized lion hunts. In ancient Egypt lions were trained to guard the pharaoh's throne or run alongside his chariot and in the Middle Ages lions could be seen at the European, Byzantine, Seljuq and Safavid courts.[23]

The Abbasid caliphs in Baghdad, likewise, impressed their guests with the lions drawn from prides that thrived in the region. A Byzantine ambassador who requested an audience with the caliph in 917 was 'led through a formation of a hundred lions, fifty on either side, each with keeper, collar and muzzle'.[24] Tame or semi-tame lions not only served as tangible symbols of the

Johanna makes love to a lion, from a copy of various works by the medieval clergyman and historian Gerald of Wales, c. 1146–1223.

ruler's dominance, but could also provide courtly diversions. The English trader John Jourdain reports that the fourth Mughal emperor, Jahangir (d. 1627), would sometimes release one amongst his courtiers 'to see if there be any soe hardie as to stand against the lion'.[25]

Some queens also kept lions at their courts. Queen T'amar of Georgia (r. 1184–1212), for example, hand-reared a boisterous lion cub. Although it was said to be bigger and fiercer than most, 'when it was brought into the palace it displayed such love and ardor for T'amar, divine in splendor, that even in double harness it could not be restrained until it laid its head on her breast and licked her face.'[26] It is inconceivable that a male monarch would be described in these terms, but Queen T'amar's relationship with her lion has some parallels.

The Roman author Aelian, who claims that domesticated lions are playful, good-tempered and obedient, describes a pet lion owned by a woman named Berenice. It shared her table, eating 'in a sober, orderly fashion' and 'would softly wash her face with its tongue and smooth away her wrinkles'.[27] Gerald of Wales, writing around 1185, tells his readers about a lion at the French court that 'used to make beastly love to a foolish woman called Johanna'.[28] The lion, which had been given as a cub to Prince Philip, frequently escaped from its cage and could not be pacified except by Johanna who would soothe him 'with a woman's tricks' and change 'his fury immediately into love'. This anecdote follows an even more salacious description of a goat that had sexual intercourse with a woman, a situation that Gerald deems 'detestable on both sides'. 'How unworthy and unspeakable!' he declares, yet he speaks, nevertheless, to the titillation or revulsion of his audience, including the learned professors of the University of Oxford, and King Henry ii of England, to whom he dedicated his book.

English monarchs first kept lions at a palace menagerie at Woodstock, near Oxford, but the animals were moved to the Tower of London by King Henry III (r. 1216–72), and lions – living emblems of the monarchy – continued to be imprisoned in the Tower until the transfer of the Royal Menagerie to the Zoological Society's Gardens in Regent's Park in 1835. Privileged visitors were admitted to the tower to see the king's animals as early as the 1420s, but it was only in later centuries that the menagerie featured on the itineraries of hundreds of visitors to London, who came especially to see the lions. Appearing in print as early as 1590, the popular phrase: 'to see the lions', meant to see the sights of London. Various superstitions grew up around the royal lions; it was said that they could predict inclement weather and identify whether or not a woman was a virgin. Since the lions' welfare was linked to that of the reigning monarch, after whom at least one lion was named, it was mandatory to keep the animals in good health and ensure that if one died it was swiftly replaced.[29] Scientists have proved that two lion skulls excavated in a moat belong to Barbary lions that lived between

A lioness and her canine companion from William Darton's *Present for a Little Boy* (1798).

the thirteenth and fifteenth centuries, but much remains to be learned about the Tower Menagerie.

The diarist Samuel Pepys, who kept an Algerian lion as a pet in the mid-1670s, made regular visits to the Tower and was particularly fond of an old lion named Crowly. As recorded in his diary, on 3 May 1662 Pepys took a group of children to see the lions. This may seem unremarkable, but it is the earliest record of ordinary, rather than royal, children visiting the Tower Menagerie.[30] In later years the lions continued to attract young people and some of the felines were immortalized in children's picture books. An engraving in a miscellany of moralistic tales entitled *A Present for a Little Boy*, published by William Darton in 1798, shows a lioness with a Mona Lisa smile lying next to a little dog. As explained in the text, the lioness developed a strong affection for the animal; she refused to eat without it, and the dog, in turn, would not leave her den. When the animal was taken away and lost, a replacement had to be found in order to console the unhappy lioness.

Stories of lions cultivating friendships with dogs and other animals are found in a wide variety of sources. Jahangir's pet lion, for example, developed a cordial relationship with a tame goat, and when the menagerie at Versailles was disbanded in 1793–4, as a result of the French Revolution, a lion and its canine companion were among the four remaining animals that were transferred to the Jardin des Plantes.[31] Captive lions are most likely to cultivate such relationships, but it has been known to happen in the wild. In December 2001 a lone lioness in Samburu National Park, Kenya, was spotted with a baby antelope, which she nurtured for sixteen days. When a male lion killed the baby oryx, the lioness adopted several others in succession, but none survived for very long, despite her solicitude. Her behaviour baffled wildlife experts, but some suggested that she probably formed these

Peter Paul Rubens, 'Lioness', c. 1614, chalk on paper, study for the painting *Daniel in the Lion's Den*.

unusual relationships to compensate for her lack of pride mates. It is probable that captive lions form similar bonds to cope with their isolation, since lions are gregarious by nature.

Over the course of the centuries many professional artists have studied lions kept in private menageries. The Flemish master Peter Paul Rubens (1577–1640), for example, sketched lions at the royal and archducal menageries in Brussels and Antwerp. Studies in chalk or oil of isolated animals were then incorporated into vast canvases with biblical or historical themes. In *Daniel in the Lions' Den* (*c.* 1614–16) the Old Testament hero is surrounded by a group of lions that look eager to devour him, but are miraculously thwarted. More spectacular is the *Lion Hunt* (1621), now in the Alte Pinakothek, Munich, which shows horsemen doing battle with lions. Despite the painting's title, Rubens'

Schelte Adamsz. Bolswert, engraving (*c.* 1630–45) after Peter Paul Rubens, *Lion Hunt* (1621).

Rembrandt van
Rijn, *Chained
lioness*, c. 1638–42,
drawing on paper.

lions are cast as the aggressors – savage forces threatening to overthrow civilization. Although he is better known for acres of female flesh, Rubens' lions had a lasting impact on future generations of artists. Marvelling at the Flemish painter's mastery of animals, the English Academician Sir Joshua Reynolds remarked: 'animals, particularly lions and horses, are so admirable, that it may be said they were never properly represented but by him'.[32]

Menageries enabled both artists and scientists to observe exotic animals at first hand. Increasing interest in animal physiognomy, pioneered in the Renaissance by the Neapolitan polymath Giovanni Battista Della Porta, who published his *De humana physiognomonia* in 1586, led to a reconsideration of the relationship between man and animals. In the seventeenth century Charles Le Brun, First Painter to King Louis XIV of France and founder of the French Royal Academy of Painting and Sculpture, took up Della Porta's ideas. Le Brun's famous series of drawings juxtaposing human and animal faces underscored the similarity between the two and implied that characteristics of the animal (e.g. the lion's 'fierce and carnivorous aspect') were shared by his human look-alike.

Charles Le Brun, 'Analogie entre le visage humain et la gueule du lion', 19th-century print after his 17th-century drawing.

Although he had little formal artistic training, George Stubbs (1724–1806), the son of an English currier, achieved popular acclaim as a painter in the mid-eighteenth century. Fascinated by the theme of a lion attacking a horse, he painted seventeen different versions of the subject. Stubbs sketched lions in private menageries, made countless drawings of horses, and performed dissections in order to understand their anatomy. Since it would not have been practical to let a lion attack his equine models, Stubbs would push a broom towards the horses to evoke an expression of terror.[33] Evidently, this method proved effective. Horace Walpole wrote a poem in his honour entitled, 'On seeing the celebrated Startled horse, painted by the inimitable Mr Stubbs', and the motif made the artist famous. But

George Stubbs, *A Horse Affrighted by a Lion*, 1777, etching.

Stubbs did not always portray the lion as a vicious predator. He also produced many canvases of serene-looking lions in idyllic settings. Ironically, these carefully rendered, sympathetic images 'of animal autonomy depended on Stubbs' observation of captive specimens'.[34] Like many of his contemporaries he was fascinated by the notion of comparative anatomy and near the end of his life he was still engaged in making comparative studies of animals as disparate as humans, tigers and chickens. Stubbs' scientific interests led him to form close ties with John Hunter (1728–1793), one of the foremost comparative anatomists of the eighteenth century.

Wishing to study animal movement and behaviour, Hunter, a Scottish surgeon, kept a well-stocked menagerie at his house in Earl's Court, London, and he collected as many animals as he possibly could, both dead and alive. Some came from travelling showmen, and others from private menageries such as Exeter 'Change in the Strand. Hunter acquired 'the liver, kidney and tongue from a lion in the Tower of London menagerie – possibly an elderly lion named Pompey which died there in 1758', and some of his other lion specimens probably came from the same source.[35] A skilled dissector, Hunter would slice open organisms and examine each component to find out how animals and human beings functioned, and his research on lions was among the earliest to reveal the inner workings of the big cats.

Hunter never missed a chance to augment his holdings, and he scoured the capital for specimens. For instance, he once charged into the bookshop of his friend George Nichol and asked him for five guineas to purchase, as he explained, 'a magnificent tiger which is now dying in Castle Street'.[36] Although this was a considerable sum, Nicol assented and the tiger was duly dissected.

Leading naturalists whom Hunter had befriended brought him creatures from distant lands. Writing to a friend in Africa, he hinted broadly, 'If a foal camel was put into a tub of spirits I should be glad. Is it possible to get a young tame lion, or indeed any other beast or bird?'[37] Many requests, veiled or otherwise, must have been answered in the affirmative. At the time of his death on 16 October 1793 Hunter had amassed a collection of 13,682 individual specimens, encompassing 500 different species.[38] Visitors to the Hunterian Museum at the Royal College of Surgeons in London, where the surviving items are exhibited to this day, may view, among other leonine remains, the toe of a lion, a lion's tongue, the upper part of a lion's oesophagus, part of the mandible of a lion or tiger, a lion's intestines and organs of digestion, and part of the rectum of a lioness.

In the following centuries dissections of animals and the proliferation of accurate anatomical drawings reinforced the idea that animals and humans were more closely linked than had been previously thought. A realistic plaster cast modelled after the flayed foreleg of a lion and bearing an uncanny resemblance

The front of a lion's tongue showing the spiny papillae, which enable it to scrape meat from bones and rid itself of ticks and flies. Specimen prepared by John Hunter.

to a human arm and hand was made by the French artist Jacques-Nicolas Brunot about 1817, and later exhibited in the Muséum national d'Histoire Naturelle, Paris. The similarity between a lion's foreleg and a human arm was also noted by the English anatomist Charles Bell who observed, 'it is very remarkable that the muscles of the arm and hand should resemble so closely the fore-extremity of the lion'.[39] Although they robbed animals of their liberty, menageries advanced artistic developments and scientific knowledge – two spheres of inquiry that often overlapped. Artists like Stubbs dissected animals and scientists like Hunter commissioned drawings to document their findings. For these individuals, and for a certain Charles Darwin, who later drew inspiration from Hunter's collection, animals held the key to the meaning of life.

Eugène Delacroix (1798–1863), one of the leading French painters of the nineteenth century, claimed 'that anatomical similarities between man and animal are proof of their deeper

Eugène Delacroix, *Head of a Lion*, n.d., pencil drawing.

affinities', and he repeatedly visited the Jardin des Plantes, Paris, with his friend, the sculptor Antoine-Louis Barye.[40] These trips to the zoo and to the adjoining Museum of Natural History and Cabinet of Comparative Anatomy laid the foundation for Delacroix's lifelong interest in lions and transformed his way of thinking. In his journal he wrote: 'Tigers, panthers, jaguars, lions, etc. Why is it that these things have stirred me so much? Can it be because I have gone outside the everyday thoughts that are my world; away from the street that is my entire universe?'[41] At feeding time, when the cats hurled themselves against the bars of their cages, Delacroix was ecstatic.

A trip to North Africa in 1832, during which he made countless sketches and watercolours, gave him a lifetime supply of themes. Delacroix's *Lion Hunt* (1861), now in the Art Institute of Chicago, shows a pair of lions grappling with North African horsemen armed with lances and sabres. In the foreground a massive male lion, with bristling mane and flashing claws, pins down one attacker, and turns to snarl at two others. Although Delacroix used captive lions as models, his slashing brushstrokes do not portray lethargic beasts in cramped surroundings but animals that snarl, glower, run, writhe and pounce.

In a note dashed off to Barye on 19 June 1829 Delacroix wrote, 'The lion is dead. Come at a gallop. It is time for us to act. A thousand friendships. Saturday.'[42] The artists were not hurrying to the lion's funeral, but to the dissection table where they studied and drew the flayed corpse. Barye, described by art critic Théophile Gautier as the 'Michelangelo of the menagerie', was especially gifted at anatomical drawing and taught the subject. Fortunately for Delacroix and Barye, but not for the exotic animals themselves, the Jardin des Plantes was acquiring new specimens on a regular basis 'due to France's expanding diplomatic and colonial activities'.[43]

Antoine-Louis Barye accentuates the animal's muscular torso in his *Standing Lioness*, 1890, bronze.

The English artist Sir Edwin Henry Landseer (1802–1873), whose portrait of Van Amburgh had delighted Queen Victoria, was a contemporary of Delacroix and shared the French artist's interest in the representation of animals. As a boy Landseer often sketched lions at the Tower of London and at Exeter 'Change, and he contributed five drawings to *Twenty Engravings of Lions, Tigers, Panthers and Leopards* (1823), written by his father John, and illustrated by his brother Thomas, who executed the engravings. At the beginning of his artistic career, when one of the lions at Exeter 'Change died, the young Landseer and an acquaintance acquired the body from the proprietor, dissected it and preserved the skin and skeleton for future study.[44]

Using local lions as models had its pitfalls, however. Edwin Landseer's painting of a dead lion, entitled *The Desert* (also known as *A Fallen Monarch*), was based on his studies of the corpse of a large male sent to him by the Secretary of the Zoological Society, London. People who viewed the painting at the

Edwin Landseer,
bronze lion at the
base of Nelson's
Column, Trafalgar
Square, London,
installed in 1867 –
photographed
around 1926.

Royal Academy Exhibition in 1849 were struck by 'its scale and grimness' and many recognized the subject. As the art historian Diana Donald writes, 'While the title and setting of the picture invite us to imagine the likely circumstances of a lion's death in the wilds of Africa, it remained fully recognisable to the public of the day as the portrayal of an actual familiar lion that had recently died in the zoo.'[45]

Landseer is best known for the bronze lions he designed for Nelson's column in Trafalgar Square, London. These took eight years to complete but were finally installed in 1867. To Landseer's dismay, when the lions were unveiled they were criticized for their lack of verisimilitude. In the late nineteenth century people were familiar with the big cats. Lions toured Britain in travelling menageries and popular journals featured engravings of lions and other exotic creatures. By this time there was no shortage of self-appointed experts willing to offer unsolicited critiques.

A generation younger than Delacroix, Barye and Landseer, the French painter Rosa Bonheur (1822–1899), a trouser-wearing, chain-smoking feminist, was the most famous *animalier* (animal artist) of her day. She too studied the animals in the Jardin des Plantes, but wished to observe lions without constraints. Tired of making repeated trips to the zoo and of renting a lion named Brutus from a certain M. Bidel, she purchased a pair of her own and had them shipped from Marseilles to her country house near Fontainebleau.[46] Nero, one of the pair, soon recognized Bonheur's voice and whenever she petted him he would stand up on his hind legs. But the bulky borders consumed alarming quantities of beef, so Bonheur reluctantly consigned them to the Jardin des Plantes. A few years later, however, she could not resist acquiring Fathma, an affectionate lioness who followed her around like a poodle. Bonheur's gift for rendering animals earned her many awards, including a gold medal at the Salon of 1848, granted to her by a jury including Delacroix, Ingres, Corot and Meissonier.[47]

By the end of the nineteenth century Sunday painters would congregate at the cages of the Jardin des Plantes, setting up their easels as close to the animals as possible. Though he drew on a range of sources, including other artworks and ephemera, that most famous amateur painter, Henri Rousseau, also made pilgrimages to the zoo. In *La Bohémienne Endormie* (The Sleeping Gypsy), painted by the enigmatic customs officer in 1897, the lion that nuzzles the gypsy woman asleep in the desert is not a threatening presence. With its velvet coat and voluptuous mane, Rousseau's lion looks like an oversized plush toy. The vigilant lion serves as a foil for the somnolent gypsy; the two share the same canvas, but they remain in their own worlds.

Technological advances brought new means for artists to engage with animals. Famous for his photographs of a galloping racehorse lifting all four hooves off the ground, Eadweard Muybridge (1830–1904) was one of the first photographers to document the lion in captivity. While the equine portraits capture the horse's dynamism, the images of a lion pacing back and forth in its cage are a sorry testament to animal locomotion. Lions can gallop faster than horses, but not in confined spaces.

For centuries lions had been locked up in cramped cages and were separated from their human admirers by sturdy bars or doors, but in the nineteenth century ideas about displaying wild animals began to change. The German animal trainer Carl Hagenbeck, who claimed to use gentler methods than his

Eadweard Muybridge, 'Time-lapse photographs of a lion pacing in its cage', from his *Animal Locomotion* (1887).

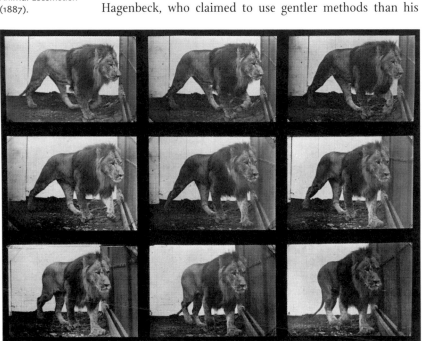

predecessors to teach circus lions tricks, also advocated giving them more freedom at the zoo. In his zoological park at Stellingen on the outskirts of Hamburg, which opened in 1907, he created an open-air enclosure for lions which mimicked their natural habitat and gave the public increased viewing access. Using discreetly placed moats to separate predator from prey, Hagenbeck showcased mixed groups of animals, creating complex landscapes or panoramas. Ducks and flamingos frolicked in the foreground of the African Panorama, while zebras, antelope and ostriches gambolled on the 'plains' behind. Lions and vultures occupied the next quadrant beneath an artificial mountain, which was surmounted by ibex and barbary sheep.[48] Photographs of Hagenbeck, some taken within the lion grotto at his zoo, show how comfortable he was with the carnivores. His favourite animal, a lioness named Triest, followed him around and would lie at his feet like a dog.[49]

Although Hagenbeck displayed affection for animals, his most lucrative business venture was the capture and sale of wild ones, including lions, whose mothers were destroyed so the cubs could be seized and tamed. As explained by Hagenbeck's biographer, Heinrich Leutemann, there was no other way to catch the big cats; 'Without exception, lions are captured as cubs after the mother has been killed, the same happens with tigers, because these animals, when caught as adults in such things as traps and pits, are too powerful and untameable, and usually die while resisting.'[50] In the space of twenty years between 1866 and 1886 Hagenbeck's animal collectors captured '1,000 lions, 400 tigers, at least 700 leopards, 1,000 bears, 800 hyenas, 300 elephants, 17 Indian, Javan and Sumatran rhinos, 9 African rhinos, at least 100,000 birds and tens of thousands of monkeys.'[51] Even if the animals survived the journey, and hundreds did not, they suffered considerable distress when uprooted from their homes and deposited in circuses and zoos.

It is no longer considered ethical to capture wild animals, but Hagenbeck's methods of displaying them in enclosures that replicated their natural habitats made a lasting impression and his innovations are reflected in modern zoos. Although freeing lions from cages may be beneficial to them, the chief beneficiaries are human visitors who want to believe that the animals, though captive, live in relative freedom and that artificial habitats are viable substitutes for the real thing.

Keeping lions captive impinges on them in fundamental ways. Captive lions, like all large wide-ranging carnivores, cannot fulfil two basic urges: to hunt and to roam.[52] Patrolling its territories and hunting is fundamental to a lion's life. Whereas captive herbivores 'can continue to spend much time eating as they would in the wild', this is not true of the big cats.[53] In addition, in circuses and zoos lions cannot lead complex social lives shared by pride mates with whom they form multiple alliances and associations. Finally, lions in captivity are forced into proximity with humans. When a lion is taken from its natural habitat and incarcerated in an artificial environment, a powerful and autonomous animal that has no natural enemies is transformed into a subordinate creature utterly dependent on its human keepers.

Animals in zoos often present a disappointing spectacle to visitors. In his seminal essay, 'Why Look at Animals', John Berger observes that exhibits rarely live up to expectations, and visitors to zoos are likely to overhear children asking, 'Where is he?' 'Why doesn't he move?' 'Is he dead?'[54] For many visitors the most dissatisfying aspects of the zoo experience are the animals' apparent indifference and listlessness. Lions, which are able to ignore even the most protracted attempts to engage them, are particularly unsatisfying in these respects. Nevertheless, visitors who assume that the lions' lethargy is induced by their captivity may be mistaken. As outlined in the previous chapter, even wild lions do very little; their lives are characterized by short bursts of energy, usually expended in hunting, followed by hours of concentrated napping. A study conducted at the Brookfield Zoo, Illinois, in 2002, revealed that lions were inactive for more than 85 per cent of the time in the spring and 90 per cent in the summer.[55]

Lazy lions may look harmless, but they retain their predatory instincts. To the consternation of visitors and staff alike, a lioness at Greater Vancouver Zoo, Canada, recently dispatched a prized exhibit, a golden eagle that flew into her enclosure, but most captive cats have few opportunities to showcase their hunting skills. Individuals who invade the lions' enclosures or cages generally do not emerge intact, and inattentive visitors who get too close risk provoking the cats. On 5 April 2000 a six-year-old boy was torn out of his father's arms and devoured by five caged circus lions in Recife, Brazil. These lions had never hunted in their entire lives, but preserved an innate ability to do so.

Not surprisingly, ditties about lions eating children have only proved popular in places where the possibility remains remote.

Examples by British and American writers include Helaire Belloc's 'Jim', Maurice Sendak's 'Pierre: A Cautionary Tale in Five Chapters and a Prologue' and Shel Silverstein's 'It's Dark in Here', which is set in a lion's stomach. Marriott Edgar's monologue 'The Lion and Albert', performed by Stanley Holloway, was immensely popular in the 1930s, and a cover version read by the British pop singer Jarvis Cocker was released in 2006.

Advocates for zoos stress that these institutions play a vital role in the understanding and preventing of diseases in animals, in education, and in the conservation of endangered species. Mission statements aside, most people who visit zoos are compelled by curiosity, nostalgia, an affinity for animals and a desire to escape the monotony of their urban lives. Lions have always drawn the biggest crowds, and a recent study shows that they still consistently rank 'as either the number one or number two species with respect to visitor interest'.[56]

Specially prepared meals with vitamin and mineral supplements have generally replaced the bloody slabs of meat thrown to zoo lions in former times, and crowds jostling to view the

A zookeeper nurses two lion cubs, Basel Zoo, c. 1908.

carnivores at feeding times are increasingly rare. Visitors are no longer permitted to dandle young lions as they once were, but the cat's charisma remains undiminished. Whether we are seduced by the lions' beauty or drawn to danger, we want to be near them.

Family posing with lions at the Hellabrunn Zoo, Munich, Germany.

3 Lion Lore and Legend

A man and a lion stand face to face, arms and forepaws entwined. Is this an image of love or enmity? Are they locked in an embrace or engaged in mortal combat? Depictions of Hercules wrestling the lion adorn a variety of objects, ranging from coins minted in Asia Minor to frescoes painted for Renaissance princes. A snow sculpture of Hercules battling the lion was modelled in Florence during the harsh winter of 1406, but a tantalizing description is all that remains of this ephemeral creation. The myth is well known: Hercules, endowed with supernatural strength, set out to conquer a lion that was terrorizing the people of Nemea. When he discovered that the lion's skin was impervious to weapons, he strangled the beast with his bare hands. Having overcome the lion, the first of twelve tasks imposed on him by his cousin Eurystheus, Hercules donned its skin. How did he skin an animal whose hide was impermeable? The problem defeated the hero until the goddess Athena told him to use one of the animal's claws as a tool.

Draped in the furry pelt, Hercules overcame all opponents and obstacles, but he was impeded when he fell in love with Omphale, Queen of Lydia, who made him comply with her every whim. Reducing the muscular hero to a figure of fun, she donned the lion skin, dressed Hercules in feminine finery, and ordered him to spin alongside her ladies-in-waiting. Robbed of his leonine attributes, Hercules was no longer the epitome of masculine power. The moral of the story is clear: lions are ferocious, but women can be just as dangerous; it is not a question of who wears the trousers but who wears the lion's skin.

We must go much further back in time to discover the earliest tale of a hero battling lions. The epic poem *Gilgamesh*, composed in Iraq about 4,000 years ago, tells how the hero, in search of immortality, travelled to the underworld and back again.

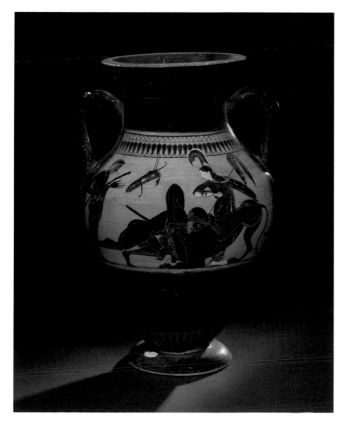

Hercules killing the Nemean lion, painting on an amphora excavated at Nola (520–510 BCE).

A hero, identified
as Gilgamesh,
overpowering a
lion, stone relief
from the throne
room of the palace
of Sargon II at
Khorsabad, eighth
century BCE.

Although Gilgamesh killed lions along the way, this is not of central importance to the narrative. In the visual arts, however, Gilgamesh is often depicted with the animal. On ancient cylinder seals and other objects, the kingly hero is shown fighting pairs of rampant lions, and a pair of monumental relief sculptures from the Palace of Sargon II at Khorsabad (northern Iraq), carved *c.* 706 BCE, portrays a muscular figure with a lion tucked under his arm like a Frenchman clutching a baguette.

Gilgamesh and Hercules set the precedent for biblical heroes, such as Samson, who tore a lion apart with his bare hands, and

David, the shepherd boy destined to become the king of Israel. David's lion-killing credentials were so impressive that he was entrusted with the job of slaying the giant Goliath, another task he performed with ease. A fable, attributed to Aesop, explains why heroes always emerge victorious. A man and a lion were walking down a road when they came upon a sculpture of a man strangling a lion. Turning to the lion, the man remarked, 'You see, we men are more powerful than you.' 'Is that so?' retorted the lion, 'If lions carved monuments, we would be the winners.'

Pet. Paul Rubenius pinxit. Franciscus vanden Wyngaerde fecit et excudit.

Frans van den Wyngaerde, etching of Samson killing the lion after Peter Paul Rubens, 17th century.

In works of art lions are represented far more frequently than lionesses. There are two reasons for this. First, lions endowed with manes are less likely to be confused with other felines. Second, lions are employed as symbols for male heroes and kings. Just as the great cat dominates other animals, the hero prevails in the hunt and in war. The idea that the hero is endowed with leonine attributes has recurred in diverse cultural contexts.

Comparisons are made, for example, between heroes and lions in Homer's *Iliad*. Covered in gore, Automedon resembles a lion that has just killed a bull, and Aias with his broad shield stands his ground against the enemy like a lioness defending her cubs. Lion similes also occur in the *Odyssey*. Penelope is too distraught to sleep because she is assailed by doubts and fears like a lion surrounded by hunters, and the Cyclops Polyphemus devours Odysseus' companions with leonine zest, 'leaving nothing, neither entrails nor flesh, marrow nor bones'. Homeric similes are complex and diverse; many evoke lions, but no two are precisely the same.[1]

Lions are worthy adversaries of noble warriors because they exemplify characteristics prized by the hero: strength, courage and fortitude. In the *Shahnama* (Book of Kings), Iran's national epic, composed by the poet Ferdowsi (935–1020 CE), several heroes engage in combat with lions. This is not surprising since the Asiatic lion (*Panthera leo persica*) was plentiful in Persia during the reign of the Sassanian kings, and common in Ferdowsi's own day.

On his long journey to Mazanderan to rescue the Shah from the demon king, the mighty hero Rostam fell asleep in a bed of reeds. Little did he know the place was a lion's lair, and when the animal returned it sprang at Rostam's horse, Rakhsh (Lightning), who was grazing nearby. Rakhsh was no ordinary horse – it was said that he could see an ant's footprints on a dark

This dynamic composition, attributed to Sultan Muhammad, is a masterpiece of Persian painting. Rustam sleeps while Rakhsh kills a lion, detail from a copy of the *Shahnama*, Tabriz, Iran, c. 1510–20.

cloth from ten miles away – and he did not shy away from the lion. Driving his front hooves into the lion's head, he seized the hapless cat between his jaws, flung it to the ground and tore it to pieces. When Rostam woke up, he was displeased. 'Rakhsh', he said, 'this was very inadvisable: who told you to fight with lions?'[2] Horses are not supposed to kill lions. Grappling with the great cats is the hero's prerogative.

Another hero of the *Shahnama*, the Sassanian prince Bahram Gur, was famous for his prowess in both the bedroom and the battlefield. One day while he was hunting, he caught sight of a lion that had pounced on a wild ass. Taking aim, he killed the pair with a single arrow. This was not for reasons of economy, but to flaunt his marksmanship. Sassanian archers were renowned for their accuracy, but it is unlikely that any, outside the realm of fantasy, could dispatch a lion in this way. Astonishing feats and death-defying ordeals are routine for heroes, however. Before Bahram Gur could win the throne, he had to battle two lions set to guard it. Defeating the beasts, the prince snatched his crown from the royal seat, demonstrating his *farr* (royal charisma).

Ferdowsi completed the *Shahnama* around 1000 CE. Approximately a hundred years later European poets and troubadours began to write down the stories of King Arthur and his knights that had been circulating in oral form for many generations. No lions were to be found in the forests of France or Britain at the time these stories were composed, although many western Europeans were introduced to exotic flora and fauna when they travelled to the Holy Land on pilgrimage or a Crusade. One chronicler of the Crusades, writing in 1184, described how a knight named Galfier de Lastours saved a lion from a dragon, and another recounted how King Louis IX of France (1214–1270) delighted in hunting the animals on horseback.[3]

Lions are featured in countless medieval books and works of art. Some are characterized as foes, others as allies. After entering the Castle of Wonders, Gawain, nephew of King Arthur, lopped off the head and forepaws of the monstrous lion set to guard it, breaking an evil spell and liberating its inhabitants. Gawain battled the lion in the chamber containing the Perilous Bed, an imposing piece of furniture, which shrieked aloud when he clambered onto it.

When they barred his approach, Lancelot of the Lake mowed down two lions that were guarding the tomb of his grandfather who had been brutally murdered. Before the lions attacked Lancelot they had to lash themselves with their tails to work up a rage. As the narrator explains, 'such is the lion's custom', because 'it will never harm man, woman or beast, until it is angry and wrathful'.[4] The idea that lions are tolerant and even-tempered is a venerable one, first advanced by classical authors, such as Pliny the Elder (23–79 CE) and Aelian (175–235 CE). Although it may seem absurd, this notion held sway for centuries.

After Lancelot had killed the lions, a hermit explained how they had come to guard the tomb. Having wounded each other while squabbling over a stag, the lions happened to lick the tomb and were healed. Blood, miraculously dripping from it, had previously cured wounded knights, and the injured lions benefited from the same sanguinary salve. The motif of the healing blood is best understood in the context of the medieval cult of relics, and of rituals centred on the Eucharist. Blood allegedly shed by various martyrs and that of Christ was venerated by medieval Christians and believed to have miraculous properties.

Happily, some legendary lions were spared the sword and they made firm friendships with people. In the *Roman de Kanor*, for example, a lion becomes the faithful companion of the emperor Cassidorus. When the latter is murdered, the lion avenges his death and transfers his allegiance to the emperor's quadruplets, entrusting the baby boys to a kindly hermit and becoming their cherished pet. A loyal lion also occurs in *Yvain, le Chevalier au Lion* (Yvain, the Knight of the Lion), written by the poet Chrétien de Troyes, whose patroness was the Countess Marie de Champagne, daughter of Eleanor of Aquitaine. As Chrétien reports, one day while Yvain was riding through the forest, he saw a lion fighting with a serpent. Slaying the reptile, Yvain rescued the lion, which padded after him, slept at his feet and defended him like a faithful dog. During the course of their adventures the bond between them grew increasingly strong. Even when imprisoned, the lion never failed to break free and come to the knight's aid, and he helped him dispatch a giant, demons and other implacable adversaries. Once, when Yvain fainted, the lion thought that he had died and was so

A lion licks the tomb of Lancelot's grandfather, from a manuscript of the *Legend of the Holy Grail*, France, c. 1470.

A lion rescues infants, from a manuscript of the *Roman de Kanor*, Paris, 14th century.

overcome with grief that he attempted suicide. Yvain reciprocated the animal's devotion. When the lion was wounded, the
knight gently lifted him on to his shield and carried him away
to convalesce.

It is not possible to summarize all of the alliances forged
between lions and legendary heroes of medieval Europe, to
tell, for example, of the cub rescued by Percival or the lion that
befriended him when he was stranded on an island. Nor can we
consider Guy of Warwick, who, like Yvain, rescued a lion from
a serpent and enjoyed the cat's company. Suffice it to say that Guy,
consumed with rage and grief when a jealous rival murdered
his lion, killed the man.

Lions dominate, but lionesses also take their place in medieval legend. In one French tale, written in the mid-thirteenth
century, a lioness abducts an infant named Octavian; she does

not devour him, but suckles him instead. Like many medieval plots, this one has more twists than a narwhal's tusk. Octavian grows up and is reunited with his human mother, but the lioness refuses to leave his side. Eventually, the trio travel to Jerusalem and near the end of the story, the lioness and Octavian rescue his long-lost brother and father from the Saracens.

Women do not generally spar with lions in medieval legends, but the princess Josiane is an exception to the rule. Sandwiched in a cave between two lions, she managed to stave them off long enough for the knight Beves of Hampton to come riding to her rescue. Having already killed Josiane's chamberlain and his horse, the lions toyed with Beves, pouncing on the knight and biting his shield. Fearing for his life, Josiane held them back, but Beves, fearing for his reputation, enjoined her to let the lions loose and finally overcame them.

Why was the heavily armed Beves almost devoured by the lions but the lightly clad Josiane left unharmed? The narrator supplies two different reasons. First, Josiane was spared because she was a virgin (something the lions were mysteriously able to ascertain). Second, the lions were unable to harm her because she was of royal blood. Best left untested, the idea that royalty are immune to lion attacks was reiterated throughout the centuries. In Shakespeare's *Henry IV Part 1* (II.iv) we read, 'The Lion will not touch the true prince', and the same notion is expressed in *The Mad Lover*, written by English playwright John Fletcher (1579–1625):

> Fetch the Numidian lion I brought over;
> If she be sprung from royal blood, the lion
> He'll do her reverence, else . . .
> He'll tear her all to pieces (IV, v).

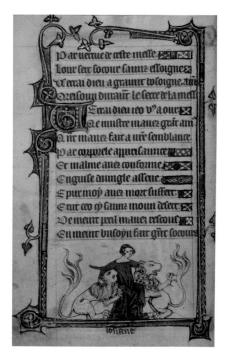

Although the lion is a fearsome predator, the animal has often been associated with the life-giving forces of sun and water. The ancient Egyptians, who observed that the annual flooding of the Nile occurred in Midsummer when the sun, for the first time, comes into conjunction with the constellation Leo, identified lions with the inundation that watered the parched earth. They also linked lions with the sun. An ivory headrest supported by the kneeling Shu, god of the air, was among the treasures discovered in the tomb of Tutankhamun. As Shu holds aloft the curved support, he is flanked by two lions lying back to back that represent two mountains between which the sun rises and sets. The head of the pharaoh, symbolizing the sun, which sinks

Josiane accosted by lions, from 'The Taymouth Hours', England, mid-14th century.

Beves of Hampton kills the lions menacing Josiane, from 'The Taymouth Hours', England, mid-14th century.

down in the evening and rises again the next morning, was thus protected each night by the lions guarding the horizon. In ancient Egypt lions lying back to back came to symbolize the concepts of yesterday (*sef*) and tomorrow (*duau*) and they mirrored the hieroglyphic sign for the horizon: the sun poised between two mountain peaks.

Several deities in the Egyptian pantheon have leonine attributes. Most have some connection with Ra, the principal sun god. Sekhmet, a vengeful goddess with the head of a lioness, is well known. Others include Shu, his sister Tefnut, Mut, Nefertem, Wadjet and Mahes. Tame lions were kept at Egyptian temple sites, notably Leontopolis (Lion City) in the Northern Nile Delta and Heliopolis, now a suburb of Cairo. The Roman author Aelian explains that the lions of Leontopolis were fed butchered animals, but sometimes a live calf was thrown to them so they could have the pleasure of killing it themselves. They were also serenaded by singers while they ate, and were embalmed and

Sehkmet and other Egyptian deities, Harris papyrus, found at Thebes, Egypt, 20th dynasty (1196–1070 BCE).

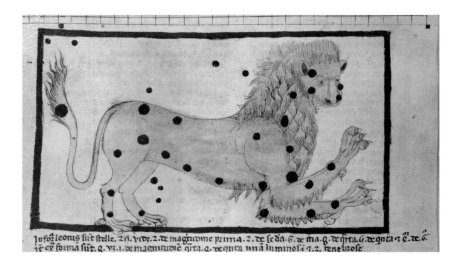

Jufoʒleonis fuͣt ſtelle. ʒ/ſ. yꝑoʒ.2.ꝺe magͤnitudine pꝛima.2.ꝺe ſeᵭa-ᵹ.ꝺe ma-ᵹ.ꝺe qͣrta.ᵹ.ꝺeqͣnta ꞇ ꞓ.ꝺe.ꞓ.
ꞇ꞊ eͣꝛ foꝛma fuͣt. ꝙ.vʒ.i.ꝺe magͤnituoic qͣtͣi.ꝗ.ꝺeqͣnta una luminoſi ꝗ.ʒ. ꞇenebꝛoſe

The constellation Leo, from an astrological manuscript, Naples, c. 1350.

entombed when they died.[5] Various sources attest to these practices, but only a single skeleton has been discovered in a tomb to date. Uncovered by a French team in 2001 at Saqqara, south of Cairo, the skeleton belongs to an unusually large male, which probably died in the first or second centuries BCE.

Over 4,000 years ago Sumerian astronomers first identified the constellation Leo. Crouched between Cancer and Virgo on the ecliptic – the zone of the sky through which the sun, moon and planets appear to move – Leo, the fifth constellation of the zodiac, still roams the night sky. In medieval Europe Leo, the celestial lion, heralded the harvest. Medieval devotional books open with calendar pages adorned with one of the twelve signs of the zodiac. Each sign is paired with a scene showing a typical activity to be accomplished during the month. Leo, the sign for July, is juxtaposed with peasants who sharpen sickles, mow meadows or thresh grain – images which celebrate the life-giving lion that rules the summer skies.

Leo, and a man harvesting grain, July calendar page from a liturgical manuscript, France, c. 1302.

Since the celestial bodies were believed to influence human health and welfare, the signs of the zodiac also appear in medieval medical manuscripts. Diagrams show a naked man with the signs of the zodiac superimposed on various parts of his body. Each sign is identified with a particular limb or organ according to principles first established by the Arabic astronomer Abu Ma'shar (d. 885). Because the lion's heart was thought to be the source of its strength, Leo presides over the chest of the 'zodiac man'.

The Nubians worshipped a lion god named Apedemak and, like the Egyptians, they identified the lion with the annual inundation. Channels carrying water from the Nile have been found during excavations at successive shrines of the lion-god built next to the royal palace in the city of Meroe, Sudan. Lions with bright streams of water gushing from their mouths could once be seen on ornamental fountains throughout the Roman Empire. In Persia, too, stone or bronze lions decorated cisterns, spouts, fountains, springs, pools and baths, protecting these highly prized resources. Water faucets in the shape of lions were also prevalent. Lions left their mark on a wide variety of objects from ponderous wellheads to miniscule flasks. Among the earliest hollow-cast metal vessels in western Europe are bronze ewers in the form of lions, used to wash the hands of wealthy diners and of priests performing liturgical rituals.

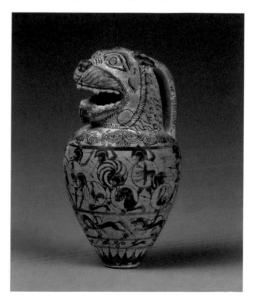

This tiny flask, designed to hold perfume, stands only 70 mm (2.75 in.) high. The Macmillan aryballos. Attributed to the Chigi Painter, Corinth, 640 BCE.

A bronze water vessel in the form of a lion, Germany, first half of the 13th century.

A lion from the Processional Way, from Robert Koldewey's *The Excavations at Babylon* (1914).

Lions have served as religious symbols for millennia, as attested by archaeological discoveries made in Iran and Iraq. Ishtar (or Inanna), a Near Eastern fertility goddess, is generally depicted standing on a lion, but sometimes the animal appears alone as her representative. Striding forward with their mouths agape, 120 lions once adorned the walls of the Processional Way leading from the Temple of Marduk to the massive Ishtar Gate built by King Nebuchadnezzar II of Babylon (605–562 BCE).

No longer a household name, the fertility goddess Cybele was venerated throughout the Mediterranean for over a thousand years. Tame lions accompanied her priests as they travelled through Greece performing exorcisms and other rituals. Cybele, who was said to control the forces of nature, is often pictured in classical art in a chariot drawn by lions. Visitors to Delphi can see her driving into battle against an army of giants on a marble frieze, carved about 525 BCE. The lion pulling Cybele's chariot attacks a giant, burying its teeth and claws in the giant's torso and testicles. The image depicts the triumph of Reason over Chaos with lions fighting on the side of Reason.

In Hindu mythology the lion can also be a force for good. Vishnu descended to earth in the form of a man-lion, Narasimha, to claw open the bowels of the tyrant Hiranyakashipu and, mounted on a white lion, the Goddess Durga defeated the buffalo demon Mahishasura. Surrounded by a high stone wall to keep out the local lions, the temple of Durga at Kankai, India, in the heart of the Gir forest, attracts up to 50,000 pilgrims a year.

Sebald Beham, 'Cybele in a Chariot drawn by two lions', pen and ink on paper, c. 1540–50.

The goddess
Durga killing
Mahishasura,
watercolour,
c. 1855–60.

Since it predates the wildlife sanctuary by several hundred years,
the shrine is destined to stay in the Asian lion's last remaining
refuge, although the steady stream of pilgrims and tourists places
an additional strain on the fragile ecosystem.

Lions play an important role in Islam. In some depictions of
Muhammad's miraculous Ascent (*mi'raj*), a lion-headed angel
is among the four celestial beings that support Allah's throne,
and in other Persian paintings Muhammad presents a ring to a
lion, a symbol of Ali, his cousin and son-in-law who is called

the 'Lion of God'. According to tradition, when Muhammad reached the seventh heaven he saw a lion and was advised by the angel Gabriel to place his ring in its mouth. Later, when Muhammad was reunited with Ali, the latter showed him the same ring. It was also said that Ali had rescued a lion from drowning in a well and that the animal followed him from that day forth. Persian legend holds that the lion rescued by Ali tried to save his son Hussein at the Battle of Karbala in 680 CE, a battle that marks the separation between Shia and Sunnis. Although the lion did not reach the battlefield in time and could only mourn the fallen Hussein, actors impersonating the animal appear in Ta'ziyeh (mourning) plays performed annually in Iran in commemoration of Hussein's death, and lions are emblazoned on the standards and banners carried by Shia Muslims during the festival of Ashura.[6]

Lions thrived in ancient Israel and are mentioned in the Hebrew Bible. Over 150 passages refer to lions directly and at least 50 others allude to them.[7] In the Book of Ezekiel, for example, the princes of Israel are likened to lions, trapped in nets and dragged into captivity, one in Egypt, the other in Babylon, so that the mountains of Israel cease to resound with their roars. In this lament the lions are characterized as man-eating tyrants, punished for their behaviour. Yet for Jews the lion is primarily a symbol of Messianic promise and redemption. As foretold by the prophets, the longed-for Messiah will be born of the royal house of David, of the Tribe of Judah, the tribe of the lion. Triumphing over darkness, the Messiah will reign with righteousness.

The importance of the lion in Jewish culture is reflected in its appearance on liturgical furnishings in the synagogue. Pairs of lions may appear on the doors of the Holy Ark (Aron ha-Kodesh), which contains the sacred scroll of the Torah, inscribed with the Pentateuch (the first five books of the Bible). Guardian lions

supporting the tablets of the law may also adorn the curtain
hanging directly in front of the Holy Ark, the mantle covering the
scroll, silver ornaments or the crown placed above it. Because
the Torah represents the presence of God among his people, the
Holy Ark is of signal importance. Worshippers pray facing the
Holy Ark, which, in the West, is usually located on the eastern
wall of the synagogue, the direction of Jerusalem.

Lions adorned liturgical furnishings in King Solomon's
Temple in Jerusalem, and his throne, crafted out of ivory and
gold, was ornamented with the animals. Two lions flanked the
armrests, and pairs of lions were arranged on each of the six steps
leading up to the throne, making it seem as if the king were
surrounded by an entire pride. Guests to the court were over-
whelmed by its splendour. In the first Book of Kings (10: 4–6) we
read, 'when the queen of Sheba saw all the wisdom of Solomon,

the house which he had built, the food on his table, the courtiers sitting round him, and his attendants standing behind in their livery . . . there was no more spirit left in her.'

According to Ethiopia's national epic, *Kebra Nagast* (*The Glory of the Kings*), Solomon seduced the Queen of Sheba and she gave birth to his son, Menelik. Entrusted with the Ark of the Covenant, containing the tablets of the law given by God to Moses, Menelik took it to Ethiopia. From that time forth, Ethiopians, graced by God's presence, identified themselves as the New Israel and the rightful heirs of Solomon. Ethiopian kings adopted the title 'Lion of Judah' and during the reign of the Ethiopian Emperor Haile Selassie I (Ras Tafari Makonnen) the lion became a symbol of liberation for the black diaspora.

Rastafarians identified Haile Selassie I (1892–1975) with Jah (God Incarnate). They saw him as a black Messiah and looked forward to the day when he would free from oppression all people of African descent. Emerging in Jamaica in the 1930s, and popularized by singers such as Bob Marley, the Rastafarian movement gained adherents who advocated personal freedom and extolled Africa. Songs, poems and paintings produced by Rastas praise the indomitable Lion of Judah, and dreadlocks, traditionally grown to signal vows of purification, resemble the lion's mane.

Surprisingly, since the lion is indigenous to Africa and is an essential element of African folklore, there is little evidence to suggest that the animal was ever worshipped outside the Nile Delta. But the predator was admired by many African peoples and continues to be revered to this day. A shaman in the Kalahari explains how he harnessed the lion's energy while dancing around a fire:

I saw a lion in it [the fire]. I trembled when I looked at it. Then the lion opened its mouth and swallowed me. The next

Narrative cycle
of Solomon and
Makeda (The
Queen of Sheba)
with Amharic
inscriptions.
Ethiopia, 20th
century.

thing I remember seeing was the lion spitting out another lion. That other lion was me. I felt the energy of the lion and roared with great authority. The power scared the people.[8]

In Africa lions were connected with kingship, an institution generally regarded as sacred. Lion skins, manes and teeth were incorporated into royal regalia and African rulers were identified with the lion through their titles, such as Mari-Jata (Lion of Mali). But the lion has never had a monopoly on kingship – many other animals were adopted as symbols of royal power. The Akan of southern Ghana, for example, privileged the leopard. In the densely forested regions in which most Akans lived lions were rare but leopards thrived. To laud an Akan chief people would compare him to a leopard, an elephant or a porcupine, rather than a lion.

In the nineteenth century, however, after British and Dutch traders had established trade links in Ghana, bringing with them their armorial crests, flags and leonine trademarks, the lion began to appear in the art of the region. The art historian Doran H. Ross demonstrates that the animal was not exploited as a decorative device in the region before the 'barrage of European-produced lion images saturated the interior of the country during the nineteenth and twentieth centuries', but that 'sometime after 1900 the lion usurped the leopard as the major feline power symbol among the Akan'.[9] Generally speaking, lions are scarce in African art and other animals prevail. Even in the lion's heartland, the savannah, where lions are celebrated in stories, proverbs and dances, they play little role in the visual arts.

Paradoxically, although lions are not native to China, the animal is well represented in Asian art. *Shishi*, female/male pairs of mythical lions, guard the entrances of Buddhist shrines and temples and once protected the tombs and residences of the

A Guardian lion in front of Qianquingmen Gate, Forbidden City, Beijing.

Chinese imperial family. Originating in China over a thousand years ago, Asian lion dances, performed today from Vancouver to Shanghai, perpetuate the idea of the guardian lion and bring the beast to life.

Lion dances vary according to region and martial arts school. Although these follow a predetermined choreography passed down by kung fu masters, lion dances also foster spontaneous displays of daring and athleticism. Two dancers inhabit the lion costume and animate the beast's fore and hindquarters. The movements are fluid, subtle, graceful; there is no distinction

Lion dancers take to the streets of London to celebrate Chinese New Year, February 2008.

between the heavy head manipulated by one performer and the cloth torso and tail animated by the other. To the sound of cymbals, drums and gongs, the lion shimmies from bakery to bank, newsagent to noodle house, travel agent to teashop, blessing the thresholds and exorcising evil before consuming its reward: a bunch of greens, suspended from a door jamb and seasoned with a generous dash of cash. For the proprietors, it is a small sum to pay in return for prosperity, longevity, health and good fortune.

Before performing lion dances, members of kung fu schools light joss sticks, make offerings of food and drink to the ancestors and pray for protection from injury – a realistic concern given the demands of the dance. A lion will sometimes mark its territory by preening and rubbing and will try to determine whether any rivals have left their scent in the neighbourhood. Battles between lions trying to devour the same vegetables were once a common feature of Chinese New Year's celebrations. But conflicts are now discouraged, and most can be avoided as long as the lions adopt deferential postures when they meet. The dancing lion of Asia is a mythical creature revered for centuries by people who embraced the foreign feline and were equally receptive to a foreign belief: Buddhism.

Although Buddhists advocate compassion towards all creatures and espouse vegetarianism, the lion, the ultimate carnivore, has symbolic resonance for them. The Buddha was not associated with the lion in his lifetime, but the connection was forged early on. Siddhartha, the Buddha, a prince born in India around 560 BCE, is known as Shakyasimha, the lion of the Shakya clan, and his teachings are referred to as Simhanda, the lion's roar. The emperor Ashoka, who united the kingdoms of India in the third century BCE, and converted to Buddhism, is famous for erecting a series of stone pillars including one surmounted by four lions standing back to back. The lion pillar, which was adopted as the national emblem of India in 1950, is located at Sarnath, near Varanasi in Uttar Pradesh, where the Buddha preached his first sermon.

In Buddhist art lions are sometimes pictured with bodhisattvas who guide people on to the path of enlightenment, or with arhats, disciples of Siddhartha. Manjusri – a bodhisattva symbolizing transcendental wisdom – is frequently depicted on the back of a lion, which serves as his throne and attribute, while his counterpart, Samantabhadra, symbolizing universal virtue, rides an elephant. The arhats were individuals who observed the

Utagawa Kunisada, 'Kabuki actors perform the *Shishi-mai* (Lion Dance)', 1851, woodblock print.

Lion capital erected by the emperor Ashoka at Sarnath, India, 3rd century BCE.

Buddha's precepts, attained enlightenment and became the focus of veneration themselves. A fourteenth-century silk scroll painted in Japan, now in the British Museum, shows an arhat, possibly Angaja, seated at the base of a waterfall attended by a lion.

The arhat Pindola Bhradvaja earned the epithet 'foremost of lion-roarers' because he quelled the doubts of monks struggling to understand the Buddha's teachings. It was said that Pindola had been a lion before assuming human form and had lived in a cave at the base of the Himalayas. Once, when Pindola was hunting, Padumuttara Buddha, who had taken compassion on him, entered his cave and began to meditate. When Pindola returned and discovered the visitor, he offered him a lotus blossom (a symbol of purity), and began to roar to ward off other animals and protect his guest. Seemingly indifferent to the noise, Padumuttara continued to meditate for several more days. He then declared that Pindola would become an arhat, and that he was destined to protect the Dharma (law). One version of the story explains that while Padumuttara Buddha was meditating, he began to levitate. Seeing the Buddha suspended in mid-air, the lion created a mound of flowers beneath him on which he could rest.[10]

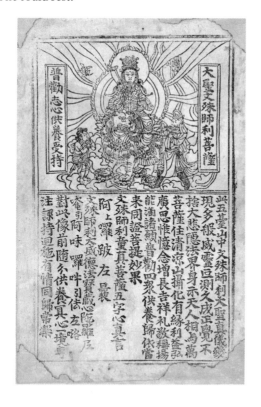

The Bodhisattva Manjusri seated on a lion, attended by a lion tamer and a small boy. China, woodcut, 10th century.

One of the
Buddha's disciples
with a lion.
Japan, silk scroll,
14th century.

Christians compared the devil to a roaring lion 'seeking whom he may devour' (1 Peter 5:8) and in medieval art the mouth of hell was sometimes shown as a lion's maw. Drawing on the Psalms, some Christian commentators equated Christ's tormentors with lions, and voracious lions, carved on thousands of corbels and capitals of medieval churches, were interpreted as symbols of unbridled passion. Moreover, a pampered nobleman riding a lion was conceived as a symbol of Pride – a vice also embodied by the lion that attacks Dante in the first canto of the *Inferno*.

Nevertheless, positive references outweigh negative ones in Christian thought. As described in the bestiary, the lioness gives birth to dead cubs, which remain inanimate until their father breathes life into them. The transformation of the cubs was interpreted as a parallel for the most important Christian miracle, God the Father's resurrection of Jesus Christ. Hymns composed for Easter celebrate the lion that emerged victorious over hell and death. 'You lion of Judah, you king, you splendid Christ, conquer the tricks of darkness with your shining light',

Lions breathe life into their cubs from a thirteenth-century English bestiary (Book of Beasts).

wrote the ninth-century poet Sedulius Scottus, and the imagery of the luminous lion, rising up in triumph, was later exploited by an Oxford don who wrote a book about a lion, a sorceress and a piece of furniture. It is easy to see why the lion with its golden coat was described as refulgent, and the positive associations of the mane, a built-in halo, made the animal an ideal Christian symbol.

The lion also acquired positive Christian connotations as the attribute of St Mark, the evangelist and Gospel writer. Christians associated Mark with a winged lion, one of four 'living creatures' seen in a vision by the prophet Ezekiel. Chroniclers claimed that St Mark's relics had been transported from Alexandria (the site

Winged lion of St Mark, patron and protector of Venice.

Lion of Judah from a theological manuscript, France, 13th century.

of the saint's martyrdom) to Venice in the ninth century, but there is little evidence to suggest that the winged lion was officially adopted as the emblem of the Venetian republic until the mid-thirteenth, a time when other Italian city-states were also laying claim to the animal.[11]

As early as 1260 Florentine officials kept lions as emblems of the state, first in an enclosure beside the baptistery and later near the city hall. Via dei Leoni (Lions' street), which lies to the east of the Palazzo Vecchio, 'is a toponymic reminder of their presence in the city'.[12] These captive lions inspired the 'Marzocco', the heraldic lion of Florence, and were a source of interest for artists. Leonardo da Vinci is said to have studied them so diligently that he was able to observe 'the structure of their eyes and the coarseness of their tongues'.[13] His fascination with the animals manifested itself in the mechanical lions he designed for the pleasure of various rulers, including King Louis XII of

France. Welcoming the king into the city of Milan in 1509, Leonardo's automaton, a lion lying above a gate, stood up and showered the monarch with lilies, which it extracted from a cavity in its chest.

A Florentine lion; the animal was adopted as a symbol of the city in the 13th century.

From the dawn of history people have conscripted lions to guard their gates. Warding off evil, stone lions have protected the thresholds of fortresses, palaces, shrines, temples, churches and public buildings, from Hittite citadels to modern office towers. Writers of medieval bestiaries claimed that lions could sleep with their eyes open, which made them ideal watchmen. But the idea of the vigilant lion goes back much farther. It has been observed, writes Aelian, 'that even when asleep the lion moves his tail, showing, as you might expect, that he is not altogether quiescent, and that, although sleep has enveloped and enfolded him, it has not subdued him as it does all other animals.'[14] Supernatural enemies were to be feared as much as

This colossal marble statue once guarded a tomb in the ancient Greek city of Knidos in Asia Minor, 2nd century BCE.

The north door of Durham cathedral with a replica of the great bronze knocker grasped by those seeking sanctuary in medieval times (the original is in the Treasury).

natural threats. The stone lions that adorn ancient tombs were not intended merely as ornaments, but to drive away evil spirits. In many cultures talismans in the form of lions were believed to offer protection from diseases and to guard against the evil eye.

Durham cathedral features a bronze doorknocker in the shape of a lion with a ring clamped between its jaws. In the Middle Ages, people who had committed a crime or stood accused of one could claim the right of sanctuary in the church by grasping the ring. The association of lions with justice was well entrenched by then. In the first century Pliny the Elder had claimed that the lion is the only animal that demonstrates compassion: *'leoni tantum ex feris clementia in supplices'* ('Among all the beasts only the lion pardons the suppliants').[15] Since Daniel had emerged unharmed from the lions' den, into which the Persian king Darius had thrown him, Jews also held that lions were capable of clemency. The biblical story even influenced Jewish legal decisions. 'If someone's husband disappeared and was known to have fallen into a den of lions, this alone was not sufficient evidence of death to allow the woman to remarry.'[16]

Similar notions regarding the mercy of lions circulated in the Muslim world well into the modern period. The British archaeologist Sir Austen Henry Layard (1817–1894) recorded the following belief while he was travelling in central Persia among tribal people known as the Lurs:

> The Lurs divide lions into Musulmans and Kafirs (infidels). The first are of a tawny or light yellow colour, the second of a dark yellow, with black mane and black hair down the middle of the back. If, they say, a man is attacked by a Musulman lion he must take off his cap and very humbly supplicate the animal in the name of Ali to have pity upon him . . . The lion will then generously spare the suppliant and depart. Such consideration, must not, however, be expected from a Kafir lion.[17]

Authors of medieval bestiaries emphasized that the lion is not easily enraged and does not oppress innocent people:

> The compassion of lions . . . is clear from innumerable examples – for they spare the prostrate; they allow such captives as they come across to go back to their own country; they prey on men rather than on women, and they do not kill children except when they are very hungry.[18]

In short, the lion served as an ideal model for human monarchs. Such positive portrayals led to the animal's incorporation in heraldic devices, notably the three Plantagenet lions of the arms of England first adopted under Richard the Lionheart

(1157–1199) and still employed today by Queen Elizabeth II. The lions on the shield represent power and majesty. But they also stand for justice, mercy and magnanimity.

Lions appear more frequently on arms than any other animal. Rampant, salient, sejant, statant, passant, combatant, guardant – this short list does not begin to describe the poses adopted by heraldic lions, which are as pliable as master yogis. The language of heraldry, derived from French, is bewildering to the uninitiated, but has an underlying logic. A 'langued lion' is a lion with its tongue sticking out, and an 'armed' one bares its claws. A 'lion queue fourchée' has a forked tail, and a 'lion defamed' no tail at all. The 'lion coward' is a rampant beast with its tail tucked between its hind legs – surely, a most inauspicious coat of arms. The charge is explained by John Guillim in his *Display of Heraldrie* (1610): 'this is termed a Lion Coward, for that in cowardly sort hee clappeth his taile between his legs, which is proper to all kind of beasts (having tailes) in case of extremity and feare.'[19] The English word 'coward' is ultimately based on the Latin *cauda* (tail), via Old French *coe*. The precise reason the tail is evoked is uncertain, but it is true that cowering animals often run away (turn tail) and, in the case of canines (but not felines), tuck the appendage between their legs.

The most famous cowardly lion appears in L. Frank Baum's *The Wonderful Wizard of Oz* (1900), described by the American author as a 'modernized fairy tale, in which . . . the heart-aches and nightmares are left out'. Anyone who has seen the film (MGM, 1939) and spent an anxious night dreaming about winged monkeys and wicked witches may dispute Baum's claim, but his portrait of the lion is indeed far from frightening.

'What makes you a coward?' asked Dorothy, looking at the great beast in wonder, for he was as big as a small horse.

A heraldic lion.

A lion spares the prostrate (above) and cringes in fear before a white cockerel (below), bestiary, England, 13th century.

'It's a mystery,' replied the Lion. 'I suppose I was born that way . . .'

'But that isn't right. The King of Beasts shouldn't be a coward,' said the Scarecrow.

'I know it,' returned the Lion, wiping a tear from his eye with the tip of his tail; 'It is my great sorrow, and makes my life very unhappy. But whenever there is danger, my heart begins to beat fast.'[20]

" You ought to be ashamed of yourself! "

The cowardly lion, illustration by W. W. Denslow from L. Frank Baum's *The Wonderful Wizard of Oz* (1900).

Pliny states that the lion is afraid of fire, the sound of creaking wheels, and roosters, and the authors of medieval bestiaries repeated these notions. Hungry lions in Homer's *Iliad* try to seize livestock, but slink off when confronted by dogs and farm hands wielding firebrands. Some nineteenth-century explorers, disappointed by the animal, also described the lion as faint-hearted. For example, the English naturalist William Burchell (1781–1863) dismissed the lion as 'an indolent skulking animal', stating, 'when men first adopted the lion as the emblem of courage . . . they were greatly mistaken'.[21] Is the lion a pusillanimous puss? Lions will avoid confrontation with humans if possible, and are sometimes frightened by innocuous things. But the designation 'coward' seems unjustified in light of the animal's characteristic resistance in face of attack.

The stereotype of the lazy lion is closer to the truth. Lions devote only a few hours each day to work and have plenty of time left for leisure. In Julian Barnes's novel *Talking it Over* (1991) Gillian, the exasperated female protagonist, compares her husband to a lion: 'Of course Oliver, like most men, is fundamentally lazy. They make one big decision and think they can spend the next few years sunning themselves like a lion on a hilltop.'[22] In Lewis Carroll's *Through the Looking-Glass*, the lion that battles the unicorn is almost too tired to bother and cannot stop yawning even when he first catches sight of Alice, a creature he has never encountered before.

> The Lion . . . looked very tired and sleepy, and his eyes were half shut. 'What's this!' he said, blinking lazily at Alice, and speaking in a deep hollow tone that sounded like the tolling of a great bell.
>
> 'Ah, what *is* it, now?' the Unicorn cried eagerly. 'You'll never guess! *I* couldn't!

The Lion looked at Alice wearily. 'Are you animal – or vegetable – or mineral?' he said, yawning at every other word.[23]

A lazy lion also occurs in Miguel de Cervantes' *Don Quixote* (1605). Eager to demonstrate his valour, Don Quixote challenges the lion to engage him in combat, but the animal is too sluggish to leave its cage. Don Quixote's companions, nonetheless, assure him that he has triumphed by default and bursting with pride he christens himself the 'Knight of the Lions'.

Cowardly lions are commonplace and lazy ones loom large, but the friendly lion is the most pervasive type of all. People have a primordial fear of predators but they also exhibit a perverse

The Lion and the Unicorn, illustration by John Tenniel from Lewis Carroll's *Through the Looking-Glass and what Alice Found There* (1871).

desire to get as close to them as possible. Tales about friendly lions appeal to us because they transcend normal boundaries. The fascination exerted by these narratives lies in the repression of the carnivore's predatory instincts. Stories about convivial herbivores, however charming, do not have the same frisson. We are captivated by the story of Elsa and the fable of Androcles because we want to believe that people and predators can establish meaningful relationships.

A Roman slave named Androcles committed a crime and fled into the desert. Seeking shelter from the sun's rays he unwittingly took refuge in a lion's den, failing to realize his mistake until the animal returned. Instead of attacking Androcles, however, the lion, which had a thorn embedded in its paw, sought his help. The two lived together for the next three years, subsisting on the meat procured by the predator, until Androcles, longing for a haircut, went in search of a barber. Unfortunately, before he could say 'short back and sides', he was apprehended by the authorities, deported to Rome and condemned to death.

By chance the lion was also captured and sent to Rome to dine on criminals, including Androcles. When the condemned man was led into the arena, he did not immediately recognize the lion, but the lion, delighted to see his friend, threw himself at his feet. Charmed by the unexpected turn of events, the emperor pardoned Androcles and freed the lion. People who met them would say, 'This is the lion that was a man's friend, this is the man who was physician to a lion.'[24]

In his play *Androcles and the Lion*, first performed in London in 1913, George Bernard Shaw recast the human protagonist as a Christian martyr, and invented a lion, which heaves 'a long sigh like wind in a trombone', 'purrs like a motor car' and 'achieves something very like a laugh'. Although Shaw tackles the subjects of religion and mortality, the tone is light-hearted.

Francis Barlow,
*Androcles and the
Lion*, etching,
1668–75.

Androcles, for example, speaks to the lion like a turn-of-the-
century nanny addressing a child: 'Oh, poor old man! Did um
get an awful thorn into um's tootsums wootsums? Has it made
um too sick to eat a nice little Christian man for um's break-
fast?'[25] Mercifully, the lion does not reply to these inanities.

While exhibiting an ability to comprehend human speech, the animal communicates in a leonine fashion with moans, growls, roars and a lashing tail.

A talking lion, is, however, featured in the *Acts of Paul* (*c.* 185–195 CE). The apocryphal text describes how the apostle chanced to meet a 'great and terrible lion' which threw itself at his feet and asked to be baptized. Paul led the animal to a nearby river and immersed him three times. 'Grace be to you!' said the lion, 'And also with you!' replied Paul. Safely back on dry land, the lion shook out his mane, shunned the overtures of a flirtatious lioness and ran off rejoicing. The story of Paul and the lion ends like that of Androcles. Paul, captured by the Roman authorities in Ephesus, was reunited with the baptized lion in the arena. The lion refused to pounce on the apostle and eventually the pair made their escape under the cover of a heavy hailstorm.[26]

There are no eyewitness accounts of Christians being thrown to lions in the Colosseum in Rome. In fact the earliest surviving

A gladiator attacked by a lion on a marble relief carving from the Temple of Artemis, Ephesus, 1st–2nd century CE.

descriptions of Christians being savaged by lions in the Colosseum were written in the fifth century, one hundred years after Christianity had been adopted as the official religion of the Roman Empire. Authors such as Suetonius and Martial do attest that 'criminals' were thrown to the lions in Roman arenas in the first century, a punishment known as *damnatio ad bestias* (condemnation to the wild beasts), but many more lions than humans were slaughtered during spectacles.[27] Captured in Africa and shipped to imperial cities, thousands of lions were forced to do battle with other beasts or gladiators, and were killed for the delectation of Roman citizens. Pompey, for example, is said to have had 600 lions and 410 leopards slaughtered on a single occasion.

Friendly lions appear in the biographies of several Christian holy men who lived in the deserts of Egypt, Syria and Palestine. We read of a hermit who plucked a handful of dates and offered them to a lion that gulped them down like a spoiled house cat, and of an old man who habitually fed lions in his cave – the Early Christian equivalent of a pensioner scattering bread for pigeons. Many of the individuals mentioned in these accounts are anonymous, but some are known to us, including St Macarius, to whom the Archangel Raphael gave two orphaned lion cubs; St Mary of Egypt, a penitent prostitute who died alone in the desert, but was buried by a monk and a solicitous lion; and St Anthony, who was aided by two lions while he was digging a grave for Paul the Hermit. Can lions dig? Anyone who has seen a lion dig a warthog out of its burrow knows that the cats can dig as well as any dog.

The name and deeds of the abbot Gerasimos have likewise been preserved for posterity. Gerasimos, who established a monastery near Jericho, once extracted a sharp reed from the paw of a lion who served him from that day forth and guarded the monks' donkey. Wrongly accused of eating the donkey, the innocent lion was eventually vindicated. Despite being unjustly blamed, the lion's love for Gerasimos never wavered, and when the abbot died he was inconsolable. 'Nothing they [the monks]

St Samuel riding a lion. Preparatory sketch for a mural painting, made in an Italian military ledger by an Ethiopian priest, 1940s.

said could appease his roaring and lamenting: for the more they tried to caress and console him the more he grieved and the louder he roared.' Heartbroken, the lion collapsed and died on the abbot's grave.[28]

Over the course of time, events from the life of Gerasimos were incorporated into the biography of St Jerome (*c.* 341–420), who translated the Bible from Hebrew and Greek into Latin. It was said that Jerome had removed the thorn from the lion's paw, that Jerome had despaired over its alleged crime, that Jerome had rejoiced to hear of its innocence, and that the lion, reduced to a tearful heap, kept vigil at the grave when his beloved Jerome died.

Due to a case of mistaken identity, Gerasimos was shunted off to one side as Western artists down through the centuries drew, painted and carved images of St Jerome and the lion. Nevertheless, Gerasimos' lion was not forgotten and was upheld as an example of heartfelt grief. Gregory of Constantinople, stressing the anguish felt by the Bulgarian monk Romylos (d. *c.* 1371), who left his master's service just before the old man died, wrote that he hurried to the grave and began to wail so bitterly that if his companion 'had not, with exhorting words, persuaded him to rise from there, he probably would have suffered like that wild beast, the lion on the grave of the blessed Gerasimos'.[29] Furthermore, on icons produced by Eastern Orthodox Christians down to the present day Gerasimos is shown alongside his loyal lion.

It may seem surprising to find this degree of anthropomorphism outside a Disney studio, but the description of Gerasimos' grief-stricken lion, written by the Syrian monk John Moschus (*c.* 550–619), is in keeping with other accounts of the desert fathers. Many biographers like John Moschus, who once lived in the Jordan Valley, were familiar with the wilderness and respected the natural world.

The stories of the desert saints reflect a new kind of relationship between humans and animals, but it is not an equal partnership. The animal is subservient and must suppress its natural instincts. Gerasimos' lion, for example, on entering the monastery, was forced to adopt the monks' vegetarian diet and was transformed from an alpha-predator into an oversized

St Jerome and the lion, from a copy of Jacobus de Voragine's *Golden Legend*, made in Paris *c.* 1480–90. Having unjustly accused the lion of killing their donkey, the monks made him serve as a beast of burden.

Hieronymus
Wierix, *Jerome
in his Study*,
c. 1566, engraving
after Albrecht
Dürer's print
of 1514.

house cat. The potential for people to abuse their authority over animals is revealed in an anecdote recounted by the twelfth-century Persian poet Farid al-Din 'Attar. An old woman carrying a bag of flour asked a Sufi saint to bear her burden. Feeling weak himself, he summoned a lion to do the job. Proudly, he asked the old lady what she would say to her friends. 'I'll tell them', she said, 'that I met a nasty show off.'[30]

Friendly animals are central to the story of Layla and Majnun, a tale of unrequited love recited throughout the Islamic world. The pair fell in love at an early age but their parents kept them apart. Consumed with love for Layla, and unable to visit her, Qays, nicknamed Majnun (Madman), lost his mind and wandered through the desert. The Persian poet Nizami (1140 – 1209), who weaves the story's simple strands into a rich tapestry, describes how Majnun befriends lions and other wild animals. These gather round the outcast, the wolf frolicking with the lamb, and the lioness suckling the orphaned fawn. 'Tigers brush against you with tenderness and affection', writes Nizami, 'lions

Majnun befriended by lions and other animals. Persian miniature attributed to Aqa Mirak from a copy of Nizami's *Khamsa*, Tabriz, c. 1539–43.

gambol with you as though they were tabby cats bought from a market stall.'[31] Majnun, who subsists on roots and berries, and eats these only sparingly, lives in peace with the animals because he does them no harm.

Benevolent lions are not confined to medieval texts. In William Blake's *The Little Girl Lost* (1789) the poem's protagonist, Lyca, wanders into the desert and falls asleep. Finding the child, a lion and a lioness convey her to their cave. The story is continued in a second poem, which tells how Lyca's parents search for her and meet the lion who promises to lead them to their daughter. Reassured by this information, as few parents would be, the couple follow the animal. Blake does not describe the reunion between Lyca and her parents, but the success of their quest is

Lucas Cranach the Elder, *Adam and Eve in the Garden of Eden*; a lop-eared lion lies in the foreground (1509, woodcut).

explicit in the poem's title: *The Little Girl Found*. Meeting the lion transforms their relationship with the wilderness and they remain there from that day forth. In Blake's words, 'the desart wild' becomes 'a garden mild'.

The Garden of Eden, described in Genesis, is a place free of violence and bloodshed. The idea of a world in which all animals live in concord has captivated people for millennia. The utopian idyll is dependent on the absence of predators; if the lion were to eat the ox the equilibrium would be destroyed. Nevertheless, rehabilitated lions, wolves and bears are essential to our conception of Paradise. A Peaceable Kingdom inhabited solely by herbivores does not excite the imagination.

Descriptions of animals by Aristotle, Pliny, Aelian and the writers of the bestiaries proved highly influential. Characteristics, however preposterous, attributed to animals in classical and medieval sources continued to be assigned well into the modern era. In the nineteenth century, for example, authors were still debating whether the lion spares the prostrate. In his discourse on the animals in the Tower Menagerie, London, published in 1829, Edward Turner Bennett writes, 'true it is that on some occasions

A lioness and her keeper from Edward Turner Bennett's *The Tower Menagerie* (1829).

the Lion has been known, in the capriciousness of his disposition, to suffer his prostrate prey to escape but little injured from his clutch'. But he argues that it would be absurd 'to conclude from such whims and freaks . . . that he is actuated by feelings of mercy, or by the natural impulse of a generous mind'.[32] The sincerity with which Bennett advances his argument suggests that he anticipates some of his readers will cling to their preconceptions of the cat's capacity for clemency.

The purpose of animal fables, from Aesop to Jean de la Fontaine, was to illuminate human virtues and vices. Nevertheless, with their simplistic portrayals of the cunning fox, the timid hare and the mighty lion, their authors perpetuated animal stereotypes. Fables generally present a pessimistic picture of society. Predators oppress herbivores, and the weak give way to the strong. In 'The Lion's Share', for example, attributed to Aesop, a lion hunting with other animals not only takes the portion to which he is entitled, but also claims the entire prize by force. In other fables, however, animals of inferior size and strength get the better of the great cat. The prevalence of the theme of the outwitted lion suggests that the lion's royal pedigree and unassailable position at the top of the food chain made him an irresistible target for satirists.

According to Aesop, a gnat once triumphed over a lion, although his victory celebration was cut short by a spider's web. And as recounted in the *Tales of Kalila and Dimna*, based on the Sanskrit *Panchatantra*, a hare, about to be eaten by a lion, also prevailed. The hare told the lion that another lion was lurking nearby wishing to steal his food. When the hare led the lion to the mouth of a well, and instructed him to look down, he could see that the animal had been speaking the truth: there, indeed, was his rival, and there was the juicy hare. Incensed, the lion lunged at his enemy and drowned, and the hare hopped away unharmed. The story of the hare and the lion is a fable-within-a fable, recounted by the jackals Kalila and Dimna, scheming courtiers of the Lion King, who goad him into killing his ally and counsellor, the bull Shanzaba. Sadly, the dim-witted lion commits the crime and is filled with remorse.

Animal fables have lost their currency in the West, where they have been sanitized and relegated to children's books. But in the late seventeenth century, when La Fontaine published his famous fables, the same sorts of stories incited the interest of a wide public who began to question their implications; 'the ascription of thought and speech to animals in fables ran counter to the mechanistic theory of animal existence proposed by Descartes and his followers'.[33] A lion with cognitive powers was a lion to be reckoned with, not an unfeeling machine.

Many people held the traditional view that God had given Adam dominion over all animals and that humans were, therefore, free to hunt, kill, beat and eat them as they saw fit, but some Enlightenment thinkers questioned man's right to rule the beasts. Scientific studies prompted new philosophical and ethical enquiries regarding man's relationship to animals. Some suggested that animals might even be superior to humans. One of the most eloquent proponents of this view was a talking lion encountered by a shipwrecked merchant on the coast of Africa.

As recounted by Bernard Mandeville in his *Fable of the Bees* (1714), the merchant 'having heard much of the generosity of lions, fell down prostrate before him with all the signs of fear and submission'.[34] The lion 'who had lately filled his belly' and had a philosophical disposition, agreed not to eat the merchant if the man could offer 'any tolerable reasons why he should not be devoured.' The merchant argued that humans were superior to animals, but the lion refuted this, saying, 'we [lions] follow the instinct of our nature. . . It is only man, mischievous man,

بوندہ یدی وہ بان اوکنده دوشندی شیر بادہ دل الك حواب خر کوشنده الذیی عظمت روت
اولدی خر کوش شیری برجاہ عیت کاربر کورردی کہ آیۂ خانه صفا دہ زایۂ صفا جین کے صو ری کا
کوستدی و یی خط طو خط اصفحۂ جین نطاری و ہ رنادا ایدردی بیت

صفا دہ شوبلہ کیت مہر منور ∴ کور بورہ ای اکا نہ یہ بکند

ایدی ای طلا خصیم نا بکا راشیو نا بکاردہ دہرایک ملایتدن و یی الیدم کورتلك
یا یتدا اورسنه خصیم اکا کو سترم شیری خر کوشی دوشنده الوب جاہده نطر لندی کوشکی
اول خر کوشك صوند بوردہ کوروب خبال ایدی اكه اول اشیی بد فعال و اول خر کورین کہ
کدہ اوجمون ایبال اول ملتلہ الحالۂ الخر کوشی قوبوب کندہ دہ لر اول بہ جاہ ہۂ آ تدی

و برایك غطیرایلہ تقیر لیتی زیا بنه زیا بنہ بمہ بہ مشا لبم ایدی خر کوش ایدی سلامنلہ دونو

that can make death a sport.' How can humans take the moral high ground? The lion is right: despite our cognitive powers, linguistic virtuosity, religious impulses and ethical pronouncements, we are gratuitous killers and dubious custodians of the natural world.

A Hare outwits a Lion, from a copy of the *Humayun-nama*, c. 1589, a Turkish version of the *Tales of Kalila und Dimna*.

4 In Pursuit of the Lion

Inexperienced lion hunters would do well to heed the advice of the British sportsman Sir Alfred Edward Pease (1857–1939), who concludes his book on lions with a chapter entitled, 'A Few Hints for Beginners'. Pease instructs hunters to clean guns and restock cartridges before setting off, to keep the lion in sight at all times and to ensure that an adequate supply of antiseptics and bandages are to hand. 'Remember', he writes, 'a large proportion of accidents are due to the failure to realize, before it has been seen, the velocity of a lion's charge. The best way to obtain an idea of it is to place yourself for a moment in front of a motor-car coming straight at you, 100 yards off, going at a speed of, say, 40 miles an hour.'[1] Whether would-be lion hunters tried this at home is unclear, but they may have considered Pease's recommendations regarding firearms and projectiles, and studied the diagrams showing the frontal and broadside targets presented by a lion. Helpful hint: the broadside target is preferable, because it offers the options of aiming for the brain, neck, spine or heart. 'A high pheasant may require as much skill to kill neatly as a charging lion', observes Pease, 'but it does not matter so much if you miss him.'[2]

Lions have rarely been hunted for their nutritional value, although some people have savoured their meat, which is said to taste like veal. Tiger skins have been more highly prized as trophies

A charging lion from Alfred E. Pease's *Travel and Sport in Africa* (1902).

than tawny lion pelts and, unlike its striped cousin, the lion has not been hunted almost to extinction because of the reputed medicinal properties of its bones and internal organs (although lion fat has been used to treat back pain and other ailments, and the heart eaten to instil courage). Moreover, despite their reputation for savagery, lions are far less interested in killing people than vice versa. Unprovoked attacks by marauding lions, while undoubtedly traumatic, are relatively infrequent. If lions prefer to be left alone and are not considered vital sources of food, medicine or adornment, what prompts people to kill them?

Hunting lions has been a royal pastime since at least 3000 BCE and in the intervening centuries, in many different places and periods, the pursuit of the lion has been deemed a royal prerogative.[3] From the ancient Egyptian pharaohs to the Mughal emperors of India, rulers who successfully battled the king of beasts demonstrated their mastery over the world at large, sending a clear message to loyal subjects and dissenters alike. Since the hunt served to reflect and legitimize political power, little was left to chance. Royal hunts were often staged or semi-staged affairs with measures introduced to reduce risks and guarantee success. Lions were often hunted within confined spaces, and a phalanx of beaters and royal huntsmen ensured that the king made the kill with a minimum of difficulty.[4]

In India hunting lions has been a royal sport for centuries, and it was one in which Jahangir (r. 1605–1627), the fourth Mughal emperor, excelled. Jahangir 'was addicted to hunting in the same way that he was addicted to alcohol and opium, and moderation of any kind was out of the question.'[5] His memoirs detail each of his hunting expeditions and enumerate his personal kills: 17,167 animals in 39 years. Lions comprise a small percentage of this tally, but the 86 killed by Jahangir are featured at the top of the list and they provided some of his most memorable moments.

Prince Khurram saves Anup Rai who had gone to the rescue of Jahanghir, painting by Balchand, c. 1640, from the *Padshahnama*.

While hunting in 1610 Jahangir and his party were surprised by a lion that bolted out of a thicket. In their haste to run away some of the huntsmen trampled the emperor, but the courtier Anup Rai stood his ground and saved Jahangir's life. Although savaged by the lion, Anup Rai survived his injuries and, in

honour of his bravery, Jahangir christened him 'Ani Rai Singh-dalan' ('Commander of troops, lion crusher'). The event is depicted in a Mughal painting, and another shows Jahangir shooting a lioness through the eye during a hunt in 1615. The emperor had been challenged to hit the target by Prince Karan of Mewar and managed to do so from his bumpy perch on the back of an elephant. 'God almighty did not allow me to be ashamed before the prince', wrote Jahangir, 'and as I had agreed, I shot her in the eye.'[6] Some Mughal lion hunts were less spontaneous. According to Francois Bernier, French physician to the sixth Mughal emperor Aurangzeb (r. 1658–1707), a tethered ass that had been force-fed opium was used to bait an Indian lion. Feeling the effects of the drug, the lion staggered around in a soporific haze before being shot by the emperor.[7]

The oldest extant hunting record for the lion takes the form of an inscription concerning the Egyptian pharaoh Amenhotep III (1390 to 1353 BCE), lord of Syria, Phoenicia, Egypt and much of the Sudan. Carved in hieroglyphics on steatite scarabs, the inscription states that Amenhotep 'brought down with his own arrows' 102 lions in the first ten years of his reign. Over 170 scarabs commemorating the achievements of Amenhotep III survive, now dispersed in various museums, and the largest number of these feature the lion hunt inscription. Curiously, his use of scarabs to record his hunting successes is unparalleled; no other Egyptian pharaoh did likewise.[8] In addition, at his mortuary temple in Thebes he installed hundreds of standing and seated figures of the lion-headed goddess Sekhmet ('She who is powerful').[9] These images reflect the religious importance of the leonine goddess to Amenhotep, the earliest ruler to have left us with a tangible record of his lion kills.

In ancient Assyria captive lions were released in designated spaces so that the king could kill large numbers in a single session.

According to the inscription on this scarab, Amenhotep III killed 102 lions in the first decade of his reign, Egypt, c. 1380 BCE.

A cuneiform inscription carved on one of the gypsum bas-relief panels commissioned by Ashurbanipal (r. 669 to 630) for the north palace at Nineveh (northern Iraq), reads: 'I am Ashurbanipal, king of hosts, king of Assyria. In my abounding, princely strength I seized a lion of the desert by his tail, and at the command of Enurta and Nergal, the gods who are my helpers, I smashed his skull with the axe in my hands.'[10] Ashurbanipal's audacious technique is depicted on the relief, and subsequent panels show other moments of the carefully choreographed lion hunt. As spectators watch from the safety of a hill, lions released from cages rush forward to attack the king and his retainers, who are mounted on horseback or ensconced in chariots. Arrows and spears are used to bring down the lions, and in one scene the king on foot, wielding a sword, seizes a powerful male by the throat and delivers the coup de grâce. Paralysed by arrows that have severed her spinal cord, a lioness drags her hindquarters along the ground, determined to face her attackers despite her injuries. Nearby another dying lion vomits blood. Spearmen

A lion is pierced by arrows, from a series of stone lion-hunt reliefs commissioned by king Ashurbanipal, Nineveh, northern Iraq, c. 645 BCE.

and dog-handlers with mastiffs guard the perimeter of the hunting arena to ensure that none of the animals escape.

A lion that declines to fight is goaded with a whip. Another, which had been left for dead, unexpectedly springs at a horse like a demented jack-in-the-box. When all of the lions have been dispatched, the king pours libations over the corpses of four sizable males in honour of Ishtar, goddess of war, to whom the lion was sacred. As this panel suggests, one of the primary purposes of the Assyrian lion hunt was to acknowledge the supernatural source of the king's power.

Discovered in the mid-nineteenth century, during British excavations in northern Iraq, these Assyrian sculptures were shipped to the British Museum, London, where they were greeted with excitement. Competition for antiquities between Western nations was intense, the sculptures were of high artistic merit and Assyria was a household name to Bible-reading Europeans. 'The marketing of Assyria in mid-century England was a multi-media affair. It soon became possible to purchase, for fine gentlemen, gold lapel studs with embossed winged Assyrian bulls and, for the ladies, bracelets and necklaces sporting royal and mythological motifs from the palace sculptures.'[11] Furthermore, the Assyrian reliefs arrived in London at a time when hunting lions had acquired a new relevance and allure. *The Illustrated London News* and other popular journals reported both archaeological discoveries and the exploits of adventurers. There was a demand for books, such as Roualeyn George Gordon Cumming's *Five Years of a Hunter's Life in the Far Interior of South Africa* (1850), which briefly outsold Dickens, and David Livingstone's *Missionary Travels and Researches in South Africa* (1857), in which he described how he had been mauled by a lion during a hunt with Tswana tribesmen in the northern Transvaal.

In the lion hunt reliefs King Ashurbanipal not only advertises his royal might, but also presents himself as the defender of

Gold bracelet with a steatite cylinder seal, designed by the firm of John Brogden, c. 1860, shows King Ashurbanipal offering libations after a successful lion hunt.

his people. His protective role is emphasized in an inscription which vividly evokes the clash between lions and pastoralists. It reads:

> Since I have succeeded to the throne of my father . . . lions have bred in mighty numbers . . . They constantly kill the livestock of the fields, and they spill the blood of men and cattle. The herdsmen and the supervisors are weeping; the families are in mourning. The misdeeds of these lions have been reported to me . . . In [the] course of my expedition I have penetrated their hiding places and destroyed their lairs.[12]

'The Missionary's Escape from the Lion', from David Livingstone's *Missionary Travels and Researches in South Africa* (1857).

The lions of Iraq, so plentiful in the Fertile Crescent in the days of Ashurbanipal, could not compete with farmers and herdsmen and were eventually eliminated. By the first decades of the twentieth century the lion had been hunted to extinction in the region.

African pastoralists living in close proximity to lions have also had to hunt them in order to protect their domestic animals. Rather than killing lions indiscriminately, African hunters identified and eliminated specific animals that had acquired a taste for livestock. Widespread taboos prevented women from participating in the hunt, and some hunters banned any mention of them lest this jeopardize the success of the enterprise.[13] Rituals, performed before and after the kill, differed from group to group, but all such activities demonstrated the hunters' respect for the lion. Esteem for the animal was also reflected in the significance assigned to the killing of the lion, which was viewed by the Maasai, for example, as a rite of passage. It is not surprising that constructions of masculinity centred on the lion hunt since it constituted the most dangerous activity apart from going to war.

Bows, and arrows whose tips had been saturated with poison, were the weapons of choice of some bushmen and of the east African Akamba, who would creep up to a lion, preferably a satiated, sleeping one, and unleash the deadly dart. To ensure success, Kalahari Bushmen recited formulas and made potent concoctions from various substances, including the genitals, eyes, fat, heart and liver of lions killed on previous occasions. This was spread on the lion's pugmarks, on weapons and on the noses of the dogs used for tracking.[14]

Hunters, including the Matabele and Zulu in the south of the continent and the Maasai, Nandi, Samburu and Lumbwa in the east, imitated the very strategies used by lions to seize their prey: stalking and striking. Trackers learned to read every sign of the lions' presence, however minute, from broken twigs and fresh deposits of excrement, to depressions in the grass or dirt. Once the lion had been uncovered the hunters would surround it, forming a ring to prevent the animal from escaping. A warrior,

Maasai herdsman
with his cattle.

sheltering behind an ox-hide shield, would then advance and
challenge the lion, and his companions would rush forward to
drive their spears into its body. Techniques varied. The Matebele
stabbed the lion with a comparatively light spear, but the Nandi
and Maasai wielded heavier ones which they threw with tremendous force in order to kill the animal as quickly as possible. How
quickly was the lion dispatched? If all went according to plan,
the lion was dead in a matter of minutes.

In 1910 the American filmmaker Carl Akeley made repeated
attempts to document Nandi warriors of East Africa spearing a
lion. But his motion picture cameras were too slow to capture the
evasive actions of fleeing lions and the flashes of the hunters'
spears. Fourteen lions and five leopards were slaughtered as
Akeley's cameras rolled, but none of the film footage was of
sufficient quality to be screened. Akeley was so disappointed
with the results that he invented the famous Akeley Camera, which
became 'the standard tool for newsreels' and was one of the
most innovative motion picture cameras on the market.[15]

Mounted on a gyroscopic base, Akeley's new camera enabled filmmakers to follow the action and take sweeping panoramic shots, making it ideal for tracking wildlife. It was also easy to reload and focus, and the viewfinder remained fixed even when the camera was tilted. Using his new camera Akeley and filmmakers Martin and Osa Johnson successfully documented a Lumbwa lion hunt in 1926 while on safari with George Eastman, president and founder of the Kodak Company. And Akeley drew on his earlier experience of the Nandi lion hunt to create a series of bronze sculptures for the American Museum of Natural History, New York (a second set was acquired by the Field Museum, Chicago).

Since African hunters, using traditional methods, drew close to the lion and were armed with a single spear, the battle was more equitable and the odds of injury much higher than when the lion was approached by a man in a car or on horseback armed with a high velocity rifle. Chants and victory dances signalled success, and the hunter to have thrown the first spear or to have grasped the lion's tail received a reward – often the lion's mane. Individuals were lauded for bravery, but the emphasis was on the communal nature of the hunt, with each man working

'Close Quarters; Tati River, November 22, 1876', from Frederick Courteney Selous's *A Hunter's Wanderings in Africa* (1895).

in cooperation to safeguard his livestock and family members. By contrast Europeans glorified the solitary encounter of big game hunter and lion. Photographs of hunters posing with their trophies perpetuated the myth of the individual single-handedly overcoming the animal.

In the 1890s, as European farmers and ranchers settled in East Africa, protecting domestic animals from leonine predators led to hunting on a large scale. Concerted efforts were made to exterminate the lion; hundreds were shot, but traps and poison were also employed. Lions were classified as vermin and their tawny hides fetched as little as £1 a piece.[16] John A. Hunter (1887–1963), a Scottish safari guide and friend of Denys Finch Hatton, single-handedly killed several hundred lions in the Ngorongoro crater. As a young man Hunter honed his marksmanship while working as a guard for the Uganda Railway. The conductor would blow the whistle twice when he saw a lion and three times for a leopard, and Hunter, shooting from the train window, would make the kill and collect the cash.[17] Settlers in Rhodesia were also encouraged to kill carnivores, and between 1903 and 1914 the government paid a bounty to hunters who killed lions, leopards and cheetahs.[18] Three decades later, in 1945, concerned to protect the cattle industry, the government of Rhodesia set out regulations for the destruction of lions, with little regard for their future.

Indian rulers placed similar bounties on various carnivores, including tigers and lions. The outcome was predictable: by the 1920s lions had been almost eliminated from India and only a small number were left in the Kathiawar peninsula. Accounts of lion hunting in India in the modern period are in short supply, especially in comparison with the countless descriptions of African expeditions. By the second half of the nineteenth century few lions on the subcontinent remained to be shot, and the

British and Indian elite trained their sights on tigers, leopards and wolves instead.

Taboos against hunting females or young animals date back to the ancient world, but in the early days of colonial expansion in East Africa, when lions seemed all too plentiful, hunters had little or no compunction in shooting pregnant females or lionesses with cubs. Female lions, unlike cheetahs or leopards, are readily distinguished from males, so hunters could have spared them, but they had no desire to do so. Recounting how he and a companion bagged two lionesses in one afternoon, the big game hunter Frederick Courteney Selous (1851–1917) coolly states: 'The one killed by my friend carried in her womb three cubs (two males and a female) that would probably have seen the light a few hours later.'[19] In his *Book of the Lion*, which he dedicated to his friend, Theodore Roosevelt, Pease describes his protracted attempts to drive a lioness out from an impenetrable bush, where she had taken refuge with her cub. He failed to shift her despite 'shouting insults at her', and setting the bush on fire, but only admitted defeat when he had exhausted his patience and ammunition.

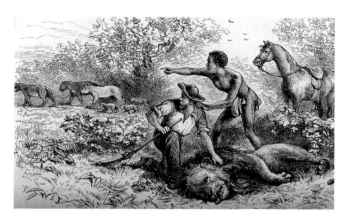

'Encounter with lions; Mashuna Land, July 16, 1880', from Frederick Courteney Selous's *A Hunter's Wanderings in Africa* (1895).

'Lions are being slaughtered like American rabbits', crowed a *New York Times* correspondent, celebrating the exploits of Paul Rainey, an American sportsman who had been invited in 1911 to hunt the 'pests' with hounds at Soysambu, the sheep farm in the Rift Valley owned by Hugh Cholmondeley, third Baron Delamere.[20]

Rainey, a playboy millionaire who owned a large property near Nairobi, had spent months training his dogs for the new challenge. As anticipated, the dogs, trained to follow the lions' spoor, cornered the big cats, while Rainey and his companions, following on horseback, delivered the lethal shot.

Moviegoers could see the carnage for themselves, since Rainey filmed much of his expedition with a state-of-the-art crank operated camera. The film, *Paul J. Rainey's African Hunt*, opened on 15 April 1912 at one of New York's premiere playhouses, Daniel Frohman's Lyceum Theatre.[21] On the big but silent screen Rainey, triumphant, poses for the cameras, grasping the ears of a lioness that had run for cover, but had been harried by the snapping hounds and finished with a bullet. *Paul J. Rainey's African Hunt*, which grossed a half-million dollars and ran for fifteen months in New York, was the most lucrative wildlife film of the period.[22] Endorsed by the American Museum of Natural History, it was touted as a motion picture that was 'simultaneously entertaining [and] educationally and morally uplifting'.[23] Trophy hunting films were among the very first motion pictures ever made. In these semi-staged or staged 'documentaries' lions and other animals 'were experienced as objects of curiosity, or as fodder for conquest'.[24]

It was variously reported that Rainey killed 12 lions in 15 days at Soysambu alone, 27 lions in 35 days, 9 lions in 35 minutes and a grand total of 74 lions.[25] Due to the high death toll, various professional hunters voiced their doubts about the ethics of hunting lions with horses and hounds, and Frederick Jackson, the British administrator in charge of issuing game licenses, concluded that it was not as sporting as hunting them on foot.[26] 'Our dogs . . . take the charge out of the king of beasts,' claimed Rainey, 'and make the pastime more pleasant.'[27] Taking the charge out of the lion was the precise problem. Was it 'fair play' for the hunter to gallop along in relative safety while the dogs distracted and weakened the animal?

Colonel Charles Jesse 'Buffalo' Jones, a Nebraskan cowboy, longed to rope the big game of Africa, and he fulfilled his dream in 1909. Jones and his fellow cowboys were faced with various logistical challenges, including the problem of removing a lasso from the neck of a giraffe, and as the final days approached, it looked like they might not fulfil their ultimate ambition: roping a lion. They had trouble locating a pride, but eventually succeeded in cornering a lone lioness. Having draped his lariat over a nearby tree, Jones lassoed her by the hind leg and hoisted her aloft where she swung 'head down like a pendulum'.[28] Although Jones released all of the other animals similarly captured, he kept

Paul Rainey's African Hunt, Soysambu, Rift Valley, Kenya, 1911.

the lioness and shipped her off to the Bronx Zoo, New York. The acquisition of a trophy – in this case a living lioness, not a truncated head or skin – was central to the experience.

Buffalo Jones' bizarre stunt is a minor footnote in the history of lion hunting, but it reveals the ambivalent attitude to animals that came to characterize big game hunting. According to field manager Guy H. Scull, who wrote an account of the expedition, after Jones had lassoed the lioness he whispered in her silky ear, 'Yes, you're a beauty. You're certainly a beauty. I guess we'll just have to take you home with us as a souvenir.' Jones' admiration for the animal made him want to keep her even though this meant destroying her way of life. Most lion hunters, both real and fictional, took their admiration further and destroyed the animal outright.

Intent on tracking a lion, Miss Mary, the protagonist of Hemingway's posthumous *True at First Light*, articulates her ambivalence towards the animal: 'He's my lion and I love him and respect him and I have to kill him.'[29] This attitude – the hunter's paradox – manifested itself especially in individuals who delighted in slaughtering animals despite professing concern for their welfare and conservation.

A prime example is the American president Theodore Roosevelt, a staunch conservationist who insisted that huge swathes of North American forest be preserved, rather than turned into planks, and that no development encroach on the Grand Canyon. In honour of his efforts and 'in recognition of his service on behalf of the preservation of species' his British admirers, including Lord Curzon, presented him with one of the finest and heaviest rifles on the market.[30] No irony was intended.

Despite his love for animals and knowledge of their behaviour, Roosevelt preferred to view them through his gun sights. In 1909–10 he spent a staggering $75,000 on safari in East Africa and bagged a total of 512 animals, including 17 lions and 9 white rhino. His contemporaries criticized his tally of rare white rhino, but his take of lions was relatively modest considering that no licence was required to hunt them and they were still classified as vermin. Professional hunters, gun-bearers and support staff accompanied the ex-president and his son Kermit, and 500 porters carried their essentials. Expert trackers were also engaged since 'Colonel Roosevelt's bulk and conversational powers somewhat precluded him from tracking.'[31]

Three scientists rounded out the hunting party in order to help Roosevelt collect suitable specimens for the Smithsonian Institute, Washington, DC, and the American Museum of Natural History in New York. Roosevelt went to Africa to shoot specimens, rather than living creatures, but the distinction was surely lost on the animals themselves. Combining sport and science, Roosevelt 'epitomized the apparent Jekyll and Hyde dilemma of the hunter-conservationist'.[32] More elaborate and widely publicized than any other African hunting expedition, Roosevelt's safari was precedent setting. A team of Associated Press reporters shadowed his every move, and photographs, film clippings and a steady stream of articles kept his African adventure in the limelight. Soon many people with disposable income wanted to go to Africa to shoot big game; 'thereafter, "safari" was an institution, very shortly a fashion'.[33]

The word 'safari', a Swahili term derived from the Arabic verb *safara* (to travel), came to mean a specific type of journey – a hunting expedition. The term conjures up men in pith helmets and ill-fitting beige clothing running around with rifles, but this picture is incomplete. As cogently argued by Edward Steinhart, the safari, 'Kenya's distinctive contribution to twentieth-century sport hunting', was not an Anglo-European phenomenon, but a 'truly cross-cultural' or 'transcultural' creation. Indigenous people had been hunting in Africa long before the arrival of Europeans and they continued to do so for decades afterwards. Steinhart writes:

> Beneath the facade of whiteness, the safari was the product of African local knowledge, African hunting values and practices and African adaptation to the imposition and negotiation of colonial rule. There can be little doubt that the contribution of African labour to the safari was as crucial as it was to other colonial enterprise.[34]

Lion hunts could not have taken place without Indian and African labour, and an absurdly high number of servants supported even the most modest expedition. Support staff for a four-man Kenyan safari in 1909, for example, consisted of 122 servants: 1 head-man, 4 gun-bearers, 4 second gun-bearers, 4 *askari* (guards), 1 cook,

Theodore Roosevelt with African hunters and a dead lion on the Athi Plains, Kenya, c. 1910.

4 tent boys, 80 porters, 4 *sais* (grooms), and finally 20 unpaid *toto* (boy porters).[35] Tasks assigned to Africans included carrying supplies and guns, cooking, making camp, cleaning, driving, guiding, and tracking and skinning animals. One head porter, a Wanyamwezi tribesman hired by William Rainsford, carried from 38 to 44 kg (85–98 lb) every day for five months.[36] Professional 'white hunters', hired to oversee the operation, ensured that their clients shot dangerous animals with the greatest ease.

The phrase 'white hunter,' coined in British East Africa in the first decade of the twentieth century, refers to professional hunters of Anglo-European origin hired to guide wealthy clients on hunting expeditions and make sure that they acquired enviable trophies. 'White hunters' were well versed in animal behaviour, and conversant to some extent with the customs and languages of their African subordinates. R. J. Cunningham, J. A. Hunter, Philip Percival, Bror Blixen and Denys Finch Hatton straddled the divide between their clients and 'untamed' Africa. Steinhart, who analyses the social dynamics of the safari, underscores the racism inherent in the hierarchical organization of such expeditions, which was reflected in the language employed by white participants. He states, 'beside the denotation of "white" to simply differentiate it from black-skinned African hunters, whose activities had by and large been outlawed and denigrated by the 1920s, the term "great white hunter" has important connotative overtones. "Great White Hunter" seems to represent a form of white supremacy over blacks.'[37]

Safari literature is replete with stories of daring hunters and lions impervious to bullets. Similar scenes are played out repeatedly with little variation in setting, action, characters and motives, leaving the reader with an acute sense of déjà vu. Both the climax (the lion's charge) and the outcome (an unwieldy tawny corpse) are predictable. Statistics and technical specifications abound. Readers are informed of the exact location of the kill, the hunter's distance from the lion(s), the number and type of bullets fired, the weight and make of the firearms employed and the number and vital statistics of the dead animals.

Even when describing fatal incidents, authors are reluctant to dispense with technical analysis, as attested by Pease's account of the untimely death of George Grey, brother of Sir Edward Grey, the British Foreign Minister: 'The late Mr George Grey, who was fatally injured in 1911 whilst hunting lions with me, hit the charging lion in the mouth and broke his jaw, at 5 yards range, with a .280 Ross copper-pointed bullet without checking

Severed head of a lion, Mozambique, photograph by C. A. Reid, 1906.

or turning the charge in the least.'[38] Pease probably included these facts in order to satisfy his aspirational readers who wanted to know precisely what and what not to do when confronted by a lion. The 'how to' tips included in the back of his book support this interpretation. But, generally speaking, the accretion of statistics by writers of safari literature may have served an additional purpose: to deflect difficult emotions. It was horrifying to witness a person being mauled by a lion, and the feelings of terror and inadequacy it evoked were often left unexpressed.

Feelings of fear, ambivalence and failure are, however, articulated in some fictional accounts. Published in 1936, Ernest Hemingway's short story *The Short Happy Life of Francis Macomber* describes the protagonist's panic and flight when charged by a wounded lion, a reaction that earned him the contempt of his wife and ultimately led to his death. Hemingway allows Francis Macomber to express his terror, and provides a sympathetic portrait of the injured lion. He writes:

> His flanks were wet and hot and flies were on the little openings the solid bullets had made in his tawny hide, and his big yellow eyes, narrowed with hate, looked straight ahead, only blinking when the pain came as he breathed, and his claws dug in the soft baked earth. All of him, pain, sickness, hatred and all of his remaining strength, was tightening into an absolute concentration for a rush.[39]

Hemingway's lion is too human to be convincing, but the change in point of view shifts the reader's sympathy from the hunter to hunted, and forces us to confront the animal's distress. By the end of the story few readers would accept without question the ideals and values of big game hunters or the stereotype of the hyper-masculine hunter, a persona Hemingway readily adopted himself.

Safari literature is self-congratulatory. This is due, in part, to the autobiographical nature of the texts, which typically take the form of memoirs, letters or journal entries. African animals are more highly valued than African people, and an undisguised *Schadenfreude* is characteristic of the genre. Describing how a wounded lion chased one of the servants in his party, Pease writes: 'I have never seen a native run in any part of Africa as this one did, doubling and turning for about 150 yards . . . the lion at his heels, and to me it appeared as if he must reach the native with his forepaws at each bound.'[40] Even while praising their black servants, 'white hunters' frequently exhibit racist attitudes and employ racist language. For example, Arnold Wienholt, in his *Story of a Lion Hunt* (1922), describes how his servants cared for him after he had been mauled by a lion: 'My three little piccaninnies stuck to me faithfully and nursed me, little savages though they were.'[41]

Several celebrated hunters were defended from lions by their gun-bearers, men who had one of the most hazardous jobs in the hunting party. Gun-bearers accompanied hunters when they approached dangerous prey, and yet they were forbidden to deploy the weapons they carried. When faced with a charging lion, acting as a portable gun cabinet was perilous at best. On a hunting expedition in 1894, when a wounded lion charged Lord Delamere and seized his leg, his Somali gun-bearer, Abdulla Ashur, threw himself at the animal and bore the brunt of its teeth and claws. Eventually, Delamere recovered his gun and drove the lion away and the two injured men spent an uncomfortable night listening to a pack of hyenas feast on the wounded beast.

African servants also saved the life of Major Percy Horace Gordon Powell-Cotton (1866–1940) who was attacked by a lion on Friday, 12 October 1906, while hunting on the banks of the Sassa River near Lake Albert Edward in the Congo. He had shot the animal several times, but it summoned the strength to charge. Pouncing on Powell-Cotton from behind, it lacerated his legs and back. Although he sustained seventeen wounds, his servants rallied to his rescue and dispatched the animal. After he had tended his injuries, the imperturbable Major had the lion photographed, gutted and carried back to camp.

Several items related to the lion attack are preserved at Powell-Cotton's estate, Quex House, on the outskirts of Birchington-on-Sea, Kent. The safari suit, complete with hat and boots, is enshrined in a glass case alongside his Lee-Metford rifle and the stick and switch wielded by the men who saved his life. An artfully arranged mirror reveals the gaping hole made by the lion in the back of his Burberry jacket and a copy of *Punch* is also displayed with a label explaining that it 'prevented the lion's claws from penetrating the Major's abdomen'. Powell-Cotton's journal contains his account of the event, and the lion's skull now rests on a shelf in the basement. But the *pièce de résistance* is the lion itself, mounted by the famous London taxidermists Rowland Ward. As if to acknowledge the lion's fighting spirit, Powell-Cotton gave it pride of place in his extensive collection of wildlife specimens – more than 500 animals are on permanent public display to this day. Rather than standing stiffly in isolation, the lion is shown in a deadly duel with a buffalo. As the lion mounts a frontal assault, teeth flashing, claws bared, the buffalo's horns come perilously close to piercing its flesh. Forever frozen in a struggle to the death, the animals exert a strange fascination. The outcome of the battle is far from certain.

It was difficult to say which African animal was the most dangerous. Buffalo and elephants were strong contenders, but a large percentage of safari-goers put lions at the top of their lists. Hunters admired lions that refused to die easily, even when perforated by a hail of bullets, and lions that put up a fight against 'fearful odds' were deemed 'sporting beasts'. The assignment of human characteristics to the lion was taken to absurd lengths. One trophy hunter, for example, professed to hold no grudge against the lion that mauled him because the animal, unwilling to be cowed, had 'behaved in true British fashion'.[42] Injuries were inevitable, but the dangers of lion hunting were sometimes exaggerated to make conquests seem more glorious, especially in the days of the motorized safari.

The introduction of motor vehicles onto the African plains in the first decades of the twentieth century had an irrevocable effect. The camera, introduced around the same time, also changed the way people conceived of big game hunting. It was no longer enough to acquire experiences; it was also necessary to obtain photographic proof. Like skins and severed heads, these celluloid trophies served to authenticate events and to give the big game hunter a glow of satisfaction long after the animals had been slaughtered. One day a man posed for the camera beside an animal he had just killed. A cursory trawl of the internet proves that the iconographic cliché conceived at that moment shows no signs of extinction. It would be difficult to exaggerate the importance of cars and cameras, both of which are constituents of the safari to this day. One of the first people to take advantage of the new technologies was, appropriately enough, George Eastman, inventor of the roll of film and founder of the Eastman Kodak Company.

In 1926, at the age of 71, George Eastman embarked on an East African safari. Journal entries and photographs, published the following year, describe Eastman's adventures, including the bagging of his first lion, a large male, which he followed in his trusty Buick until he was close enough to take aim. The kill was presumably made after Eastman had disembarked from the vehicle, but the petrol-propelled chase, which enabled him to approach the lion safely and speedily, introduced a new element into an age-old practice. A lioness and a pair of cubs eluded Eastman and his fellow hunters, but they managed to stuff the heavy carcass of the lion into the backseat of the car to transport it back to camp.

A photograph published by Eastman, bearing the caption, 'G. E.'s first lion', shows the head of the animal spilling out of the open door of the vehicle. Its cheek rests on a gun bearer's thigh, and Eastman holds aloft one of its stiff paws like the raised fist of a wounded prizefighter. But the victory is the hunter's, not the lion's, and the photograph, centred on the Buick, documents the beginning of a phenomenon that would have a detrimental impact on the African landscape. Philip Percival, among others, was later to criticize hunting with motor vehicles, but there was no turning back. Although cars often broke down and got stuck in the mud they made lion hunting more efficient than ever. In the first place, cars served as effective camouflage for the hunter. Despite the noise and fumes lions did not perceive cars as threats and did not flee from them. Second, as demonstrated by Eastman's photographs, cars made transporting trophies relatively easy. Servants remained vital to the safari, but automobiles eventually supplanted the great teams of porters used to transport goods. Third, cars enabled hunters to move quickly from one place to another. Having successfully chased and shot a cheetah,

'G. E.'s first lion', from George Eastman's *Chronicles of an African Trip* (1927).

Eastman and his companions were alerted that other members of their party had discovered a lion. Throwing the half-skinned cheetah in the back of their vehicle they took off in pursuit of the bigger prize. Finally, cars were also convenient places to display recently acquired trophies. Another photograph in Eastman's book, bearing the caption, 'A Morning's Hunt', depicts a skinned lion draped over the bonnet of an automobile like a furry figurehead, and the decapitated heads of an impala, an eland and a topi projecting from the front fender.

Trophy hunting was, and remains, a competitive sport and a quest for perfection. Unlike lions, which often hunt the most vulnerable prey, the impala with a limp or the immature zebra, human hunters strive to kill the biggest and the best. Eastman was no different, but he was pragmatic enough to make some exceptions. Fearing that he might not get a second chance, for example, he shot an inferior rhino because he needed the skin to cover a table in his new library.[43]

Aware of the decreasing number of lions, Eastman observed that if the situation continued, Africans would lose the art of lion spearing.[44] Instead of refraining from shooting them, however, Eastman – who killed five himself – took comfort in the thought that his fellow-hunters, the American filmmakers Martin and Osa Johnson, had documented local lion hunts and obtained a record for posterity.

By the mid-1930s various game laws had been passed in East Africa. Females of all species were protected, as were animals within 500 yards (450m) of water holes and salt licks. Hunting at night was no longer permitted, hunting with dogs was banned and the practice of shooting animals from automobiles was outlawed.[45] Regulations were difficult to enforce and often broken. Professional hunters such as Philip Percival lobbied for laws to protect African fauna and formed societies to promote ethical

'A Morning's Hunt', photograph by Martin Johnson, from George Eastman's *Chronicles of an African Trip* (1927).

hunting practices, such as the East African Professional Hunters' Association, established in Nairobi in 1934, one of the earliest and most influential organizations.[46] Since members of the association were in the business of killing animals and had wealthy clients to please, some people doubted that they would uphold the laws themselves.[47] But the fact that legislation was passed reveals a growing awareness that African wildlife was not an unlimited resource.

The idea that it was necessary to shoot lions and other carnivores to preserve prey species held sway into the twentieth century. Operating on this assumption the first game wardens appointed to oversee African reserves and national parks shot hundreds of lions, leopards, hyenas and wild dogs. George Schaller's seminal study of Serengeti lions showed that this was misguided. In *The Serengeti Lion: A Study of Predator-Prey Relations* (1972) Schaller argued that it was not necessary to cull lions to preserve other species, but that the animals, if left alone, could regulate themselves. The lion's appeal to tourists also helped to conserve it. Visitors to National Parks wanted to see lions, the more, the better. No longer regarded as a pest, the lion acquired celebrity status. Instead of killing lions rangers began to protect them.

Little was done to preserve the Asian lion until the animal was eradicated from all of India apart from the Kathiawar peninsula. In 1879, wishing to save them, Mahbatkhanji II, the sixth nawab of Junagadh, whose territory encompassed the lions' last bastion, issued 'an interdict . . . against the destruction of lions in the Gir forest', and his successors, likewise, tried to ensure the lions' survival.[48] While they permitted elite guests to shoot some of the great cats, they controlled the sport and shot only a modest number or none themselves. Alarmed by rumours of dwindling numbers of lions, the British Viceroy Lord Curzon famously declined to shoot one on a state visit in 1900, and he encouraged British officers to follow his example – a call which fell on deaf ears. But despite the demand for lion trophies the nawab of Junagadh's measures prevented indiscriminate slaughter.

Following the example of his predecessors, the ninth and last nawab, Mahbatkhanji III, guarded the lions jealously. In 1921 he wrote to a British official, complaining that a neighbouring ruler was poaching them, stating, 'It is an unquestionable fact that the house of the lion is the Gir forest and equally unquestioned that the forest is my ancestral property.' Forwarding the nawab's letter to his superiors, government agent E. Maconochie confessed, 'Personally I am inclined to regard the lion as a dangerous anachronism, who does a lot of damage and wastes a great deal of my time and I should contemplate his final departure with equanimity.'[49] Thankfully, not everyone shared his view. British officials in Bombay conceded that they would 'regret the complete disappearance of the lion', and although the animals were hunted freely in 1947–8, during the political upheaval that followed independence and partition, the new leaders of India deemed the Gir lions a national treasure. In 1952 the lion was adopted as the national animal of India, a position it held for twenty years until it was deposed by the tiger. More importantly, in 1965, the Gir Forest National Park Wildlife Sanctuary was established to ensure that the region would remain 'the house of the lion'.

5 Golden Remnant

Lying immobile on the Serengeti Plain or gliding between the trees of the Gir forest, lions merge with their surroundings, seeming to disappear. This is a trick of light, of camouflage, but it gives us pause for thought. What if lions were to vanish forever? More lions have been killed in the last two and a half centuries than in any previous period. Over-hunting and colonial expansion had a devastating impact in the nineteenth century, but the greatest threat to the survival of the species today is unchecked human population growth. Instead of maintaining a respectful distance from the great cats we are encroaching on their habitats and pushing them off the planet. In 1951 the population of India was estimated at 361 million.[1] It is now approximately 1.1 billion. India's largest city, Mumbai, had 7 million residents in 1975. Today it is home to 19 million.[2] The teeming metropolis is only two days drive from the Gir Sanctuary, and over 400,000 people live on the park's boundaries. How can 300 lions possibly compete?

On the Red List of threatened species compiled by the International Union for Conservation of Nature (IUCN), the Asian lion (*Panthera leo persica*) is classified as 'Endangered', and the African lion (*P. leo*), 'Vulnerable'. A recent survey estimated that roughly 23,000 lions remain in all of Africa.[3] Lions are still plentiful in East and South Africa, but elsewhere their populations are declining. In Africa, as in India, the greatest threat to the lion is

Camouflaged lioness.

human expansion and range reduction. Between 1975 and 1995, for example, mechanized farming around the Masai Mara National Reserve increased from 4,875 ha to 47,600 ha.[4] African lions, whose kingdoms once stretched across the vast continent, now 'live in islands among a sea of humanity'.[5]

Natural history programmes perpetuate the myth that Africa is an uninhabited wilderness, but the unedited footage presents a more complicated picture. In many places lions and people lead an uneasy co-existence. Large carnivores need room to roam. With people crowding in on every side, sub-adult males evicted from their natal prides may find that they have nowhere to go, and no possibility of taking over a pride and fathering cubs. If lions cannot migrate they cannot disperse their genes and give rise to the next generation of healthy individuals. Leaving the protective boundaries of natural parks and reserves, lions encroach on human settlements and prey on domestic animals. Kills frequently follow a predictable cycle. Domestic herbivores compete for food with wild ones. If resources are scarce, wild herbivores decline, and lions are apt to feed on livestock instead.

In 1995, in a single district in Kenya, lions, leopards and hyenas destroyed 202 cattle and 993 sheep.[6] Currently, in parts of Africa, poisons sold as pesticides are being deployed to kill carnivores. Granules of the toxic chemicals are sprinkled on the carcasses of decoy animals and when lions and other animals feast on these, the poison attacks their central nervous systems, causing paralysis and death. Scientists monitoring lions in Kenya report that at least 75 lions have been poisoned in the last five to six years by pastoralists seeking to protect their livestock, and the actual death toll is suspected to be much higher. Only a ban on substances capable of causing such carnage will prevent more killings.

Hemmed in by farmers and herdsmen, the Asian lions of Gir also supplement their diet with the buffalo or cattle tended by local villagers, and the lions' lives are fraught with danger. Some stray into fields and drown in open wells. Others are electrocuted by fences. Still others are struck by cars or trains.

Lions that have lost their natural reticence and acquired a taste for human flesh pose severe problems. In some parts of

An apt reminder.

Tanzania, which is home to the largest lion population in Africa, ordinary activities like hoeing a field or going to the toilet can end in a lethal attack. Men working outside are especially at risk, but lions also silence the laughter of children and pull people from their beds. Between January 1990 and September 2004 lions killed 563 people in Tanzania and injured at least 308.[7] The people of Gir have suffered similar tragedies. Chilling anecdotes and angry scars are vivid reminders that a forest filled with lions is a dangerous place to raise buffalo. Many Maldharis (local pastoralists) have been relocated to other sites, but others have declined to leave their ancestral homes within the park boundaries. Lions living or straying outside the park also cause havoc. One hundred and ninety-three attacks on humans were reported between 1978 and 1991, and 28 fatalities.[8] Lions were more antagonistic towards people in the aftermath of a drought, which reduced the numbers of wild animals on which they usually preyed. Simple yet effective measures such as building stronger enclosures to protect cattle have been shown to reduce predation by lions, and killing repeat offenders has helped reduce confrontations.

Man-eating lions are rare, but they never fail to make headlines. Yet the truth is that people are far more destructive than lions. Lions are poisoned or caught in poachers' snares. Those suspected of killing livestock, or predicted to do so, are summarily slaughtered. Animals that prey on people are, of course, destroyed, although this process is rarely as straightforward as it sounds. Innocent lions are often shot by mistake in the quest to find the real killer.

Canned hunting – the practice of breeding animals in captivity and releasing them in fenced enclosures so that hunters can shoot them with relative ease – is an abhorrent but lucrative industry estimated to account for up to 95 per cent of all lions killed in South Africa. The government has made moves to ban the practice but comprehensive legislation, forbidding all such killing, has yet to be passed.

Trophy hunting continues to grow in popularity. Statistics show that 'approximately 18,500 foreign hunting clients now visit sub-Saharan Africa annually, compared with 8,000 in 1990'.[9] Although hunters are not as numerous as tourists they spend more money per person, generating a total of us$201 million per year. Various arguments have been advanced in favour of regulated lion hunting. Compared to tourism, trophy hunting, if conducted in an ethical manner, 'has potentially lower environmental impact'.[10] On land that is not favourable for agriculture local people can generate income by managing private reserves. And tracts of land set aside for trophy hunting provide new habitats for lions. Hunting has found support in surprising quarters. Laurence Frank, a wildlife biologist at the University of California, Berkeley, who thinks hunting can function as a conservation tool, says:

> I don't understand the motivation for trophy hunting, [but] it's enormously lucrative. It's a huge industry and it requires huge landscapes full of life in order to function. According to a recent model, it takes a population of something like 200 lions to produce three trophy males a year, so a small off-take of older animals that aren't critical to the population can make the entire ecosystem pay for itself. In areas that have no real agricultural value, which is true of an awful lot of Africa, you could have livestock destroying habitat, or you could make some money out of having rich fat old men shooting animals.[11]

Lion expert Craig Packer also believes that hunting can further conservation initiatives. How could shooting lions possibly preserve them? First, it is important to note that Packer and Frank are not advocating indiscriminate hunting of lions, but a selective harvest of mature males that are at least six years of age. It is relatively easy to determine a lion's age because the tip of the nose becomes darker as the lion grows older. Statistical models indicate that killing these animals is likely to cause the least disruption to a pride since mature males will have already disseminated their genes and raised at least one batch of cubs to adulthood.[12] Killing older males rather than young ones or females would allow healthy populations of lions to be maintained. The thousands of dollars earned by selling licences to trophy hunters could be invested in local economies and conservation projects. If the lion is seen as an asset, rather than a liability, people are much more inclined to preserve it. Paradoxically, hunting bans in Kenya (1977 to present), Tanzania (1973 to 1978) and

A satiated male.

Zambia (2000 to 2003) have 'resulted in an accelerated loss of wildlife due to the removal of incentives for conservation'.[13]

A critically endangered Asian lioness.

Many people find repugnant the idea of killing animals for sport, and will not be convinced that hunters and guides will bother to distinguish between lions with pink or black noses. 'It presupposes that hunters behave in an ethical way', says Andy Loveridge, a conservationist at the University of Oxford, 'Some are respectable but some are complete cowboys.'[14] People with guns, however, are not the greatest threat to lions. The greatest threats are silent and insidious, and they strike when least expected.

On 3 February 1994 a group of tourists enjoying a hot air balloon ride over the Serengeti spotted a male lion that was acting strangely. Gripped by convulsions, he twitched and collapsed. He died later that night, disoriented and alone, the first victim of a mysterious disease. Within a year it had killed a thousand lions – one third of the total population – and it left hundreds of others with permanent neurological damage. Travelling north beyond the borders of Tanzania, the epidemic spread into Kenya and lodged in lions in the Masai Mara. What was this devastating scourge and where had it come from? The answers were only too obvious: the cats were dying of canine distemper spread by the 30,000 domestic dogs that were living on the park's periphery. Scientists believe that jackals, bat-eared foxes or hyenas, scrounging on the outskirts of local villages, picked up the virus and transmitted it to the lions.

Like celebrities pursued by paparazzi, lions grow accustomed to unsolicited attention.

Lions in the Serengeti are spread across 20,000 km. Gir lions inhabit an area ten times smaller and they are even more susceptible to disease because they have a very limited gene pool. If a similar epidemic were to occur in India, the Asian lion could be

wiped out. Recognizing that it would be expedient to move ten to fifteen Gir lions to another reserve, wildlife experts selected the Kuno-Palpur sanctuary, 800 km (500 miles) to the northeast in the state of Madhya Pradesh. Twenty-three villages were relocated to make way for the lions, an expensive and complicated process. Despite two decades of discussion, however, the lions remain at Gir. The problem is political rather than practical. Although the relocation programme has the support of the federal government and the Wildlife Institute of India, it has met with fierce resistance in the lions' present home, the state of Gujarat. Gujaratis do not want to relinquish their hold on the lions. They do not want to share their lions with the citizens of Madhya Pradesh. Divyabhanusinh, a Gujarati who nonetheless advocates the transfer of some Gir lions to the new habitat writes, 'In all this, it is forgotten that lions, like any other species, belong to no one but themselves, and should be protected for their own sake. They in turn make no claim that the human beings of Gujarat belong to them.'[15]

Tourists can be equally self-absorbed. When people plan safaris or hunting expeditions, they expect the lions to be there, awaiting their arrival, rolling out the red carpet, posing for their cameras or entering their gun sights. Lions do not sign model releases or enjoy rights of privacy. Permission to photograph them is not only impossible to obtain, but is also 'considered inapplicable and irrelevant'.[16] The impact on lions has not been fully evaluated, but 'cheetahs in Amboseli National Park, Kenya, discontinued hunts, abandoned kills and even were driven to infanticide when surrounded and disturbed by tourists'.[17] Even if tourists have a less destructive impact on lions, it is arrogant to assume that the great cats welcome our constant presence. Lions are sensitive creatures, not cash cows to be milked for profit.

The next generation.

Although natural history films, viewed by millions, raise awareness of environmental issues, filmmaking is invasive. Competing film crews take a proprietary interest in lions and stalk them like paparazzi eager to obtain the most sensational footage. Lions largely ignore safari vehicles, but 'the same cannot be said for their prey'.[18]

The 'Big Five' – elephant, buffalo, rhinoceros, leopard and lion – are big business. When political violence erupted in Kenya at the beginning of 2008 and foreign nationals cancelled their holidays, the country lost £150 million in potential revenues in only three months. Income derived from wildlife tourism, if distributed equitably, can foster local support for lion conservation programmes. Unfortunately, fair distribution of resources remains an elusive goal and poverty stands in the way of conservation. Unless local people have a stake in wildlife they will have 'no reason not to acquiesce in poaching and positive reason to engage in the practice themselves'.[19] Drawing on data obtained in Tanzania, an international team has recently observed, 'conservation attempts to sustain viable populations of African lions place the lives and livelihoods of rural people at risk in one of the poorest countries in the world'.[20] Plans for preserving the lion must encompass the interests of people living near them, not just those of bureaucrats, businessmen, non-governmental organizations and conservationists.

Ever since the first lion left its pugmarks on the planet, the great cats have crossed paths with people. They have inspired antipathy and fear, but they have also won our respect and admiration. No other animal has had such an enduring symbolic resonance, or has been evoked so consistently in myth and metaphor. But the stereotype of the king of beasts – of an invincible animal whose reign will never cease – is dangerously misleading. As a Persian poet once wrote: 'Even lions that terrify the world, and dragons, cannot escape the snare of fate.'[21] We cannot take for granted the golden remnant, the lions of the twenty-first century, born into an increasingly complex world.

Timeline of the Lion

3.5 MYA	1.87–1.7 MYA	320,000–190,000 years ago	300,000 years ago
Date of the first fossil record of a lion-like member of the *Panthera* family discovered at Laetoli in Tanzania	Date of the first definite fossil record of a lion, discovered in Olduvai, Bed 1, Tanzania	Common ancestor of the modern lion may have emerged in sub-Saharan Africa	Lions settle in North America after crossing Beringia, a large unglaciated landmass connecting Siberia and Alaska

c. 1390 BCE	*c.* 645 BCE	3rd century BCE	1st–4th century CE
Oldest extant hunting record for the lion carved in hieroglyphics on steatite scarabs during the reign of the Egyptian pharaoh Amenhotep III	Assyrian reliefs from the palace at Nineveh show king Ashurbanipal hunting lions. Earlier lion hunt reliefs were carved *c.* 850 BCE at Nimrud for king Ashurnasirpal	Ashoka, emperor of India, erects a pillar surmounted by lions at Sarnath, near Varanasi in Uttar Pradesh. The sculpture is adopted as the national emblem of India in 1950	Thousands of lions are killed in Roman arenas during games and gladiatorial contests

1839	1860s	1867	1893	19th–mid-20th cen
Isaac Van Amburgh (*d.* 1865), the American 'brute tamer', entertains Queen Victoria and her British subjects. He is one of the first performers to stick his head inside a lion's mouth	The Cape Lion of South Africa (*Panthera leo melanochaita*) is hunted to extinction	Edwin Henry Landseer's lions are installed at the base of Nelson's Column, Trafalgar Square, London	Lions trained by Carl Hagenbeck, a zookeeper and circus owner from Hamburg, Germany, entertain thousands of people at the Chicago World's Fair	Lions are eradicate from Pakistan, mo India, much of Sou Africa, Turkey, Syr Iran, Iraq, Tunisia Algeria and Moro

100,000 years ago	c. 30,000 BCE	10,000 years ago	c. 3000 BCE
The Asian subspecies (*Panthera leo persica*) emerges	Early humans paint lions on the walls of Chauvet Cave in southeastern France. In Germany, they carve lion-human hybrids from mammoth ivory	Extinction of lions in America, Europe and Northern Eurasia	Sumerian astronomers identify and name the constellation Leo

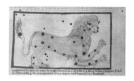

1st century	12th century	13th century	15th century
Pliny the Elder notes that lions are merciful but prone to halitosis. The fable of *Androcles and the Lion* is composed. The Roman poet Ovid popularizes the story of Hercules who kills the Nemean lion	Three lions passant first appear on the arms of England during the reign of Richard the Lionheart. The French poet Chrétien de Troyes writes *Yvain, le Chevalier au Lion*	A winged lion (attribute of St Mark) is officially adopted as the symbol of the Republic of Venice. A lion also serves as the emblem of the city-state of Florence. In England, during the reign of King Henry III, lions are housed in the Tower of London	Leonardo da Vinci designs and builds several mechanical lions. None survive

1909–10	c. 1920	1960	1994	2009
Theodore Roosevelt goes on safari in East Africa. He spends $75,000 and bags 512 animals, including 17 lions	The Barbary Lion (*Panthera leo leo*), once ranging from Morocco to Egypt, is hunted to extinction	Elsa becomes a celebrity when Joy Adamson publishes *Born Free: A Lioness of Two Worlds*. The movie *Born Free* (Columbia Pictures, 1966) soon follows	*The Lion King* (Walt Disney) wins popular acclaim. Canine distemper epidemic kills 1,000 lions – one-third of the total population – in the Serengeti-Masai Mara	One in four mammals, including lions, is threatened with extinction

References

INTRODUCTION

1 Nobuyuki Yamaguchi, Alan Cooper, Lars Werdelin and David W. Macdonald, 'Evolution of the Mane and Group-living in the Lion (*Panthera leo*): A Review', *Journal of the Zoological Society of London*, CCLXIII (2004), p. 331. See also Lars Werdelin and Margaret E. Lewis, 'Plio-Pleistocene Carnivora of Eastern Africa: Species Richness and Turnover Patterns', *Zoological Journal of the Linnean Society*, CXLIV/2 (2005), p. 129.

2 Bruce D. Patterson, *The Lions of Tsavo: Exploring the Legacy of Africa's Notorious Man-Eaters* (New York and London, 2004), p. 104.

3 J. W. Franks, A. J. Sutcliffe, M. P. Kerney and G. R. Coope, 'Haunt of Elephant and Rhinoceros: The Trafalgar Square of 100,000 Years Ago – New Discoveries', *The Illustrated London News* (14 June 1958), pp. 1011–13.

4 Chris Stringer, *Homo Britannicus: The Incredible Story of Human Life in Britain* (London, 2006), pp. 6, 67–77.

5 Yamaguchi et al., 'Evolution of the Mane and Group-living', p. 334.

6 Jean-Marie Chauvet, Eliette Brunel Deschamps and Christian Hillaire, *Dawn of Art: The Chauvet Cave* (New York, 1996). See also Jean Clottes and Marc Azéma, *Les félins de la grotte Chauvet* (Paris, 2005).

7 Nicholas J. Conard, 'Palaeolithic Ivory Sculptures from South-western Germany and the Origins of Figurative Art', *Nature*, CDXXVI, 18/25 (December 2003), pp. 830–32.

8 For an account of the discovery written by the palaeontologist who excavated the mummy, see R. Dale Guthrie, *Frozen Fauna of*

the Mammoth Steppe: The Story of Blue Babe (Chicago, IL, 1989).

9 Alan Turner and Mauricio Antón, *The Big Cats and Their Fossil Relatives: An Illustrated Guide to their Evolution and Natural History* (New York, 1997), pp. 10–11.

10 Marina Sotnikova and Pavel Nikolskiy, 'Systematic Position of the Cave Lion *Panthera spelaea* (Goldfuss) Based on Cranial and Dental Characters', *Quaternary International*, CXLII–CXLIII (2006), p. 226.

11 Yamaguchi et al., 'Evolution of the Mane and Group-living', p. 338.

12 Stephen J. O'Brien, *Tears of the Cheetah: The Genetic Secrets of Our Animal Ancestors* (New York, 2003), p. 47.

13 Frederick Courteney Selous, *A Hunter's Wanderings in Africa: Being a Narrative of Nine Years Spent Amongst the Game of the Far Interior of South Africa* (London, 1881), pp. 257–8.

14 H. Bauer and S. Van Der Merwe, 'Inventory of Free-Ranging Lions *Panthera leo* in Africa', *Oryx*, XXXVIII/1 (2004), p. 26.

15 Ross Barnett, Nobuyuki Yamaguchi, Ian Barnes and Alan Cooper, 'The Origin, Current Diversity and Future Conservation of the Modern Lion (*Panthera leo*)', *Proceedings of the Royal Society of Biological Sciences*, CCLXXIII (2006), p. 2120.

16 Pieter Kat and Chris Harvey, *Prides: The Lions of Moremi* (Washington, DC, 2000), p. 137.

17 John L. Gittleman, ed., *Carnivore Behavior, Ecology, and Evolution* (Ithaca, NY, 1989), p. vii.

18 Jonathan and Angela Scot, *Big Cat Diary: Lion* (London, 2002), pp. 20–22.

1 LIONS AT LARGE

1 Ludwig Wittgenstein, *Philosophische Untersuchungen* (*Philosophical Investigations*), trans. G.E.M. Anscombe (Oxford, 1953), p. 223.

2 George B. Schaller, *The Serengeti Lion: A Study of Predator–Prey Relations* (Chicago and London, 1972), p. 9.

3 H. S. Singh, 'Population Dynamics, Group Structure and Natural Dispersal of the Asiatic Lion (*Panthera leo persica*)', *Journal of the*

Bombay Natural History Society, xciv/1 (1997), pp. 65–70.

4 Paul J. Funston, Michael G. L. Mills, Philip R. K. Richardson and Albert S. van Jaarsveld, 'Reduced Dispersal and Opportunistic Territory Acquisition in Male Lions (*Panthera leo*)', *Journal of Zoology*, cclix (2003), pp. 131–42.

5 Judith A. Rudnai, *The Social Life of the Lion: A Study of the Behaviour of Wild Lions in the Nairobi National Park, Kenya* (Lancaster, 1973), p. 5.

6 Schaller, *Serengeti Lion*, p. 93.

7 G. Peters, 'Purring and Similar Vocalizations in Mammals', *Mammal Review*, xxxii/4 (2002), pp. 245–71, and G. E. Weissengruber, G. Forstenpointner, G. Peters, A. Kübber-Heiss and W. T. Fitch, 'Hyoid Apparatus and Pharynx in the Lion (*Panthera leo*), Jaguar (*Panthera onca*), Tiger (*Panthera tigris*), Cheetah (*Acinonyx jubatus*) and Domestic Cat (*Felis silvestris f. catus*)', *Journal of Anatomy*, cci/3 (2002), pp. 195–209.

8 Karen McComb, Craig Packer and Anne Pusey, 'Roaring and Numerical Assessment in Contests Between Groups of Female Lions, *Panthera leo*', *Animal Behaviour*, xlvii (1994), pp. 379–87, and Jon Grinnell and Karen McComb, 'Maternal Grouping as a Defense against Infanticide by Males: Evidence from Field Playback Experiments on African Lions', *Behavioral Ecology*, vii/1 (1996), pp. 55–9.

9 Craig Packer, *Into Africa* (Chicago, il, 1994), p. 69.

10 Pieter Kat and Chris Harvey, *Prides: The Lions of Moremi* (Washington, dc, 2000), p. 19.

11 Packer, *Into Africa*, p. 118.

12 Bruce D. Patterson, *The Lions of Tsavo: Exploring the Legacy of Africa's Notorious Man-Eaters* (New York and London, 2004), p. 134.

13 Rudnai, *Social Life of the Lion*, p. 45.

14 Jonathan and Angela Scot, *Big Cat Diary: Lion* (London, 2002), p. 42.

15 George Schaller, *Golden Shadows, Flying Hooves* (London, 1973), p. 78.

16 Kat and Harvey, *Prides*, p. 37.

17 George B. Schaller, *Serengeti: A Kingdom of Predators* (New York, 1972), pp. 66–7.

18 Ibid., p. 67.

19 Packer, *Into Africa*, p. 100.

20 Schaller, *Serengeti Lion*, pp. 191–2.

21 Richard Despard Estes, *The Behaviour Guide to African Mammals* (Berkeley, CA, and London, 1991), p. 371; Grinnell and McComb, 'Maternal Grouping as a Defense', p. 55; Luis A. Ebensperger, 'Strategies and Counterstrategies to Infanticide in Mammals', *Biological Reviews*, LXXIII (1998), pp. 321–46.

22 Schaller observed that lactating lionesses sometimes mate, Schaller, *Serengeti Lion*, p. 178, and Kate Nicholls, who conducted research on the reproductive cycles of lions in the Okavango Delta, Botswana, was able to determine when the lionesses came into oestrus by testing hormone levels in faecal samples; Kat and Harvey, *Prides*, pp. 34 and 79–81.

23 Grinnell and McComb, 'Maternal Grouping as a Defense', p. 55.

24 For Kruger, see Bruce Patterson, *The Lions of Tsavo* (New York, 2004), p. 125; for the Okavango, see Kat and Harvey, *Prides*, p. 98; for Asian lions, see Divyabhanusinh, *The Story of Asia's Lions* (Mumbai, 2005), p. 19; for the Kalahari, see Clive Walker, *Signs of the Wild: A Field Guide to the Tracks and Signs of the Mammals of Southern Africa*, 5th edn (Cape Town, 1996), p. 94.

25 Dereck and Beverly Joubert, *Hunting with the Moon* (Washington, DC, 1997), and Dereck Joubert, 'Hunting Behaviour of Lions (*Panthera leo*) on Elephants (*Loxodonta africana*) in the Chobe National Park, Botswana', *African Journal of Ecology*, XLIV (2006), pp. 279–81.

26 Kat and Harvey, *Prides*, p. 93.

27 H. H. Berry, 'Namibia's Seal-Eating Lions in Danger', *Cat News*, XIV (1991), p. 10, and 'Last Skeleton Coast Lions Killed by Farmers', *Cat News*, XV (1991), pp. 7–8.

28 Schaller, *Serengeti Lion*, 136.

29 Packer, *Into Africa*, 67.

30 Alfred Edward Pease, *The Book of the Lion* (London, 1913), p. 193.

31 Oliver P. Höner, Bettina Wachter, Marion L. East and Heribert Hofer, 'The Response of Spotted Hyenas to Long-Term Changes

in Prey Populations: Functional Response and Interspecific Klepto-parasitism', *Journal of Animal Ecology*, LXXI (2002), pp. 236–46, and Martina Trinkel and Gerald Kastberger, 'Competitive Interactions between Spotted Hyenas and Lions in the Etosha National Park, Namibia', *African Journal of Ecology*, XLIII (2005), pp. 220–24.

32 Hans Kruuk, *Hunter and Hunted: Relationships between Carnivores and People* (Cambridge, 2002), p. 106.

33 P. E. Stander, 'Cooperative Hunting in Lions: The Role of the Individual', *Behavioural Ecology and Sociobiology*, XXIX (1992), pp. 445–54.

34 Schaller, *Golden Shadows*, p. 65.

35 Kruuk, *Hunter and Hunted*, p. 27.

36 Schaller, *Serengeti Lion*, p. 360.

37 Patterson, *Lions of Tsavo*, p. 152.

38 Divyabhanusinh, *Asia's Lions*, pp. 19 and 226.

39 Peyton M. West and Craig Packer, 'Sexual Selection, Temperature, and the Lion's Mane', *Science*, CCXCVII (23 August 2002), p. 1342.

40 Bruce D. Patterson, 'On the Nature and Significance of Variability in Lions (*Panthera leo*)', *Evolutionary Biology*, XXXIV (2007), p. 57.

41 C.A.W. Guggisberg, *Simba* (Cape Town, 1961), p. 96.

42 Rudnai, *Social Life of the Lion*, p. 57.

43 Kat and Harvey, *Prides*, p. 27.

2 CAPTIVE CATS

1 Caroline Cass, *Joy Adamson: Behind The Mask*, 2nd edn (London, 2000), p. 130.

2 Joy Adamson, *Born Free: A Lioness of Two Worlds* (London, 1960), p. 56.

3 Cass, *Joy Adamson*, p. 146.

4 George Schaller, *Golden Shadows, Flying Hooves* (London, 1974), p. 184.

5 Stephen Duffy, 'Landseer and the Lion-tamer: The "Portrait of Mr Van Amburgh" at Yale', *British Art Journal*, III/3 (2002), p. 25.

6 William Johnson, *The Rose-Tinted Menagerie* (London, 1990), p. 50.

7   Diana Donald, *Picturing Animals in Britain, 1750–1850* (New Haven, CT, and London, 2007), p. 192.

8   'In a Circus Menagerie: Peculiarities of the Animals and their Prices', *New York Times* (30 April 1882).

9   'Trainers of Wild Animals in Fights for Life', *New York Times* (25 July 1909).

10 'A Troublesome Lion', *New York Times* (6 February 1881).

11 Nigel Rothfels, *Savages and Beasts: The Birth of the Modern Zoo* (Baltimore, MD, and London, 2002), pp. 160–61. See also Edward P. Alexander, *Museum Masters: Their Museums and their Influence* (Walnut Creek, CA, 1983), pp. 311–40.

12 'Claire Heliot – Most Daring of Lion Tamers', *New York Times* (29 October 1905).

13 Frank Norris, Joseph R. McElrath and Douglas K. Burgess, *The Apprenticeship Writings of Frank Norris, 1896–1898* (Darby, PA, 1996), p. 284.

14 'Lions Jealous of Dietrich: Actress's Theory of Why Her Pets Killed Vaudeville Man', *New York Times* (23 June 1914). On Adgie, see also Tiny Kline and Janet M. Davis, *Circus Queen and Tinker Bell: The Memoir of Tiny Kline* (Champaign, IL, 2008), pp. 164–6.

15 'For Vaudeville's Patrons', *New York Times* (2 August 1914).

16 Frank C. Bostock, *The Training of Wild Animals* (London and New York, 1903), p. 81.

17 Patricia Bourne, *Thank You, I Prefer Lions* (London, 1956), p. 63.

18 Alex Kerr, *No Bars Between* (London, 1957), p. 113.

19 Ibid., p. 93.

20 Bourne, *Thank You, I Prefer Lions*, p. 96.

21 John Stokes, '"Lion Griefs": The Wild Animal Act as Theatre', *New Theatre Quarterly*, XX (2004), p. 146.

22 At www.sonypictures.com/classics/fastcheap/index-s.html (accessed 5 October 2009).

23 Thomas T. Allsen, *The Royal Hunt in Eurasian History* (Philadelphia, PA, 2006), p. 150.

24 Ibid., p. 150.

25 Ibid., p. 151.

26 Ibid., p. 150.

27 Aelian, *On the Characteristics of Animals*, trans. A. F. Scholfield, 3 vols (Cambridge, MA, 1959), I, p. 335.

28 For this and subsequent quotations, see Gerald of Wales, *The History and Topography of Ireland*, trans. John J. O'Meara, revd edn (London, 1982), pp. 75–6.

29 Daniel Hahn, *The Tower Menagerie: The Amazing True Story of the Royal Collection of Wild Beasts* (London, 2003), pp. 105–6.

30 Ibid., pp. 117–18.

31 Richard W. Burkhardt Jr, 'A Man and His Menagerie', *Natural History*, CX/1 (February 2001), pp. 62–9.

32 Joshua Reynolds, *The Works of Sir Joshua Reynolds*, 3 vols (London, 1809), II, p. 422.

33 Donald, *Picturing Animals in Britain*, p. 73.

34 Ibid., p. 165.

35 Wendy Moore, *The Knife Man: The Extraordinary Life and Times of John Hunter, Father of Modern Surgery* (London, 2005), p. 177.

36 Ibid., p. 210.

37 Ibid., p. 351.

38 Jessie Dobson, 'John Hunter's Animals', *Journal of the History of Medicine*, XVII (1962), p. 479.

39 Eve Twose Kliman, 'Delacroix's Lions and Tigers: A Link between Man and Nature,' *Art Bulletin*, LXIV/3 (1982), p. 460, n.80.

40 Ibid., p. 457.

41 *The Journal of Eugène Delacroix*, trans. Lucy Norton (London, 1951), p. 56.

42 William R. Johnston and Simon Kelly, *Untamed: The Art of Antoine-Louis Barye* (Baltimore, MD, 2006), p. 6.

43 Ibid., p. 5.

44 Duffy, 'Landseer and the Lion-tamer', p. 25.

45 Donald, *Picturing Animals in Britain*, p. 197.

46 Anna Klumpke, *Rosa Bonheur: The Artist's (Auto)biography* (Ann Arbor, MI, 2001), p. 183.

47 Dore Ashton and Denise Browne Hare, *Rosa Bonheur: A Life and a Legend* (New York, 1981), p. 191. Less well known today than his

fellow jurors, Jean-Louis-Ernest Meissonier (1815–1891) specialized in military subjects, including portraits of Napoleon.

48 Harro Strehlow, 'Zoological Gardens of Western Europe', in *Zoo and Aquarium History: Ancient Animal Collections to Zoological Gardens*, ed. Vernon N. Kisling, Jr (Boca Raton, FL, 2000), p. 103.

49 Rothfels, *Savages and Beasts*, p. 73. A stone sculpture of the lioness adorns his memorial at the Hagenbeck Tierpark, Hamburg, which is owned by his descendants, and a recumbent Triest guards his grave at the Hauptfriedhof Ohlsdorf cemetery.

50 Heinrich Leutemann, *Lebensbeschreibung des Thierhändlers Carl Hagenbeck* (Hamburg, 1887), p. 29, as quoted by Rothfels, *Savages and Beasts*, p. 62.

51 Johnson, *Rose-Tinted Menagerie*, p. 52.

52 Ros Clubb and Georgia Mason, 'Captivity Effects on Wide-Ranging Carnivores', *Nature*, CDXXV (2 October 2003), pp. 473–4.

53 Stephen St C. Bostock, *Zoo and Animal Rights: The Ethics of Keeping Animals* (New York, 1993), p. 70.

54 John Berger, *About Looking* (London, 1980), p. 21.

55 Susan W. Margulis, Catalina Hoyos and Meegan Anderson, 'Effect of Felid Activity on Zoo Visitor Interest', *Zoo Biology*, XXII (2003), p. 597.

56 Ibid., p. 597.

3 LION LORE AND LEGEND

1 Steven H. Lonsdale, *Creatures of Speech: Lion, Herding and Hunting Similes in the Iliad* (Stuttgart, 1990).

2 Ehsan Yar-Shater, Dick Davis and Stuart Cary Welch, *The Lion and the Throne: Stories from the Shahnameh of Ferdowsi* (Washington, DC, 1998), p. 171. For a complete English translation of the *Shahnameh* see Abolqasem Ferdowsi, *Shahnameh: The Persian Book of Kings*, trans. Dick Davis (New York, 2006).

3 Velma Bourgeois Richmond, *The Legend of Guy of Warwick* (New York and London, 1996), p. 34, and Margaret Renée Bryers Shaw, trans., *Joinville and Villehardouin: Chronicles of the Crusades*

(London, 1963), p. 289.

4 Norris J. Lacy, ed., *Lancelot–Grail: The Old French Arthurian Vulgate and Post-Vulgate in Translation*, vol. III, trans. Roberta L. Krueger, William W. Kibler and Carleton W. Carroll (New York and London, 1995), p. 238.

5 Aelian, *On the Characteristics of Animals*, trans. A. F. Scholfield, 3 vols (Cambridge, MA, 1959), III, p. 19.

6 Parviz Tanavoli, *Lion Rugs: The Lion in the Art and Culture of Iran* (Basel, 1985), pp. 24–8.

7 Brent A. Strawn, *What Is Stronger than a Lion? Leonine Image and Metaphor in the Hebrew Bible and the Ancient Near East* (Göttingen, 2005), pp. 26–7 and 327–74.

8 Bradford Keeney, ed., *Ropes to God: Experiencing the Bushman Spiritual Universe* (Philadelphia, PA, 2003), p. 53.

9 Doran H. Ross, 'The Heraldic Lion in Akan Art: A Study of Motif Assimilation in Southern Ghana', *Metropolitan Museum Journal*, XVI (1981), pp. 170 and 178.

10 John S. Strong, 'The Legend of the Lion-Roarer: A Study of the Buddhist Arhat Pindola Bharadvaja', *Numen*, XXVI/1 (1979), pp. 68–70.

11 Bianca Maria Scarfi, ed., *The Lion of Venice: Studies and Research on the Bronze Statue in the Piazzetta* (Venice, 1990), p. 33, n. 5.

12 Adrian W. B. Randolph, 'Il Marzocco: Lionizing the Florentine State', in *Coming About... A Festschrift for John Shearman*, ed. Lars R. Jones and Louisa C. Matthews (Cambridge, MA, 2001), p. 11.

13 Paul Barolsky, *Why Mona Lisa Smiles and Other Tales by Vasari* (University Park, PA, 1991), p. 61. See also Jill Burke, 'Meaning and Crisis in the Early Sixteenth Century: Interpreting Leonardo's Lion', *Oxford Art Journal*, XXIX/1 (2006), pp. 77–91.

14 Aelian, *On the Characteristics of Animals*, I, p. 331.

15 Pliny the Elder, *Natural History*, trans. H. Rackham, W.H.S. Jones and D. E. Eichholz, 10 vols (Cambridge, MA, 1938–62), III, p. 36.

16 Alan Unterman, *Dictionary of Jewish Lore and Legend* (London, 1997), p. 120.

17 Austen Henry Layard, *Early Adventures in Persia, Susiana, and*

Babylonia, 2 vols (Piscataway, NJ, 2003), I, pp. 444–5.

18 T. H. White, *The Bestiary: A Book of Beasts* (New York, 1960), p. 9.

19 John Guillim, *A Display of Heraldrie* (London, 1610), p. 184.

20 L. Frank Baum, *The Wonderful Wizard of Oz* (London, 1987), pp. 68–9.

21 E. T. Bennett, *The Tower Menagerie* (London, 1829), p. 19.

22 Julian Barnes, *Talking it Over* (London, 1992), p. 247.

23 Lewis Carroll, *Through the Looking-Glass and What Alice Found There* (London, 1948), pp. 138–9.

24 Aulus Gellius, *Attic Nights*, 3 vols, trans. C. Rolfe (Cambridge, MA, 1984), I, pp. 421–7. See also Aelian, *On the Characteristics of Animals*, II, pp. 167–9.

25 George Bernard Shaw, *Androcles and the Lion* (Harmondsworth, 1949), p. 114; previous quotes at 111 and 149.

26 Tamás Adamik, 'The Baptized Lion in the Acts of Paul', in *The Apocryphal Acts of Paul and Thecla*, ed. Jan N. Bremmer (Kampen, 1996), p. 64.

27 Keith Hopkins and Mary Beard, *The Colosseum* (London, 2005), pp. 103–6 and 164–6. See also Boris A. Paschke, 'The Roman *ad bestias* Execution as a Possible Historical Background for 1 Peter 5.8', *Journal for the Study of the New Testament*, XXVIII/4 (2006), pp. 489–500.

28 John Moschus, *Pratum Spirituale*, in *Patrologia Graeca*, vol. LXXXVII, ed. J.-P. Migne, cols 2965–9. For an English translation see Helen Waddell, *Beasts and Saints* (London, 1934), pp. 25–9.

29 Mark Bartusis, Khalifa Ben Nasser, and Angelikie E. Laiou, 'Days and Deeds of a Hesychast Saint: A Translation of the Greek Life of Saint Romylos', *Byzantine Studies / Etudes Byzantines*, IX/1 (1982), pp. 24–47.

30 'Attar, *Tadhkirat al-awliya*, ed. Muhmmad Este'lami (Tehran, 2003), pp. 179–80, cited by John Renard, *Friends of God: Islamic Images of Piety, Commitment, and Servanthood* (Berkeley and Los Angeles, CA, 2008), p. 97.

31 Nizami, *Layla and Majnun*, ed. Colin Turner (London, 1997), p. 175.

32 Bennett, *The Tower Menagerie*, pp. 19–20.

33 Diana Donald, *Picturing Animals in Britain, 1750–1850*

(New Haven, CT, and London, 2007), p. 115.

34 For this and subsequent quotations, see Bernard Mandeville, *The Fable of the Bees, or Private Vices, Publick Benefits*, ed. Irwin Primer (New York, 1962), pp. 115–17.

4 IN PURSUIT OF THE LION

1 Alfred Edward Pease, *The Book of the Lion* (London, 1913), p. 255.
2 Ibid., p. 75.
3 Julian Reade, *Assyrian Sculpture* (London, 1983), p. 53.
4 Thomas T. Allsen, *The Royal Hunt in Eurasian History* (Philadelphia, PA, 2006), p. 86.
5 Ellison B. Findly, 'Jahangir's Vow of Non-Violence', *Journal of the American Oriental Society*, CVII/2 (1987), p. 247.
6 Divyabhanusinh, *The Story of Asia's Lions* (Mumbai, 2005), pp. 104–5.
7 Allsen, *Royal Hunt*, p. 88.
8 Ernest A. Wallis Budge, *Tutankhamen: Amenism, Atenism and Egyptian Monotheism* (London, 1923), p. 70.
9 Many of these statues, moved by later rulers, have been discovered at the temple of Mut (Isheru) south of the great temple of Amun at Karnak.
10 Francis Klingender, *Animals in Art and Thought to the End of the Middle Ages* (London, 1971), p. 61.
11 Steven W. Holloway, 'Biblical Assyria and Other Anxieties in the British Empire', *Journal of Religion and Society*, III (2001), p. 5.
12 C.A.W. Guggisberg , *Simba: The Life of the Lion* (Cape Town, 1961), pp. 158–9.
13 I. Schapera, 'A Native Lion Hunt in the Kalahari Desert', *Man*, XXXII (December 1932), p. 278.
14 Ibid., pp. 278–9.
15 Mark Alvey, 'The Cinema as Taxidermy: Carl Akeley and the Preservative Obsession', *Framework: The Journal of Cinema and Media*, XLVIII/1 (Spring 2007), p. 23.
16 Edward I. Steinhart, *Black Poachers, White Hunters: A Social*

History of Hunting in Colonial Kenya (Oxford, 2006), pp. 161–2, and Bartle Bull, *Safari: A Chronicle of Adventure* (London, 1988), p. 197, who cites the price paid for lion skins in East Africa in 1910.

17 Bull, *Safari*, p. 204.

18 James C. Murombedzi, 'Pre-Colonial and Colonial Conservation Practices in Southern Africa and their Legacy Today', IUCN, *The World Conservation Union* (2003), pp. 9, 11 and 14.

19 Frederick Courteney Selous, *A Hunter's Wanderings in Africa: Being a Narrative of Nine Years Spent Amongst the Game of the Far Interior of South Africa* (London, 1881), p. 258.

20 'Rainey's Big Lion Hunt', *New York Times* (23 August 1911).

21 Gregg Mitman, *Reel Nature: America's Romance with Wildlife on Films* (Cambridge, MA, 1999), p. 19.

22 Derek Bousé, *Wildlife Films* (Philadelphia, PA, 2000), p. 47.

23 Mitman, *Reel Nature*, p. 19.

24 Bousé, *Wildlife Films*, p. 125.

25 Ibid., p. 48.

26 Sir Frederick Jackson, *Early Days in East Africa* (London, 1930), pp. 382–3.

27 'Rainey's Big Lion Hunt'.

28 For a detailed description of events, see Guy H. Scull, 'Lassoing Wild Animals in Africa', *Everybody's Magazine* XXIII/3–5 (1910), pp. 309–22, 526–38, 609–21.

29 Ernest Hemingway, *True at First Light* (London, 2004), p. 106.

30 Bull, *Safari*, p. 166.

31 Ibid., p. 180.

32 Ibid., p. 160.

33 Kenneth M. Cameron, *Into Africa: The Story of the East African Safari* (London, 1990), p. 49.

34 Steinhart, *Black Poachers, White Hunters*, pp. 2 and 211.

35 The safari was made by the American reporter John McCutcheon and three companions who were following Roosevelt's expedition. It was organized by Newland, Tarlton & Company. See William K. Storey, 'Big Cats and Imperialism: Lion and Tiger Hunting in Kenya and Northern India, 1898–1930', *Journal of World History*,

II/2 (Fall 1991), p. 157.

36 William S. Rainsford, *The Land of the Lion* (London, 1909), p. 146.

37 Steinhart, *Black Poachers, White Hunters*, p. 132.

38 Pease, *The Book of the Lion*, p. 43, n. 1.

39 Ernest Hemingway, *The Complete Short Stories of Ernest Hemingway: The Finca Vigia Edition* (New York, 1987), p. 16.

40 Pease, *The Book of the Lion*, p. 221.

41 Arnold Wienholt, *The Story of a Lion Hunt* (London and New York, 1922), p. 82.

42 Ibid., p. 76.

43 George Eastman, *Chronicles of an African Trip* (Rochester, NY, 1927), p. 63.

44 Ibid., pp. 72–3.

45 Brian Herne, *White Hunters: The Golden Age of African Safaris* (New York, 1999), p. 168.

46 Ibid., pp. 167–8.

47 Steinhart, *Black Poachers, White Hunters*, p. 136.

48 Divyabhanusinh, *Asia's Lions*, p. 138.

49 Ibid., pp. 159–60.

5 GOLDEN REMNANT

1 Divyabhanusinh, *The Story of Asia's Lions* (Mumbai, 2005), p. 217.

2 At http://news.bbc.co.uk/1/hi/business/7261458.stm (accessed 24 April 2009).

3 H. Bauer and S. Van Der Merwe, 'Inventory of Free-Ranging Lions *Panthera leo* in Africa', *Oryx*, XXXVIII/1 (2004), pp. 26–31, pp. 27 and 30.

4 K. Homewood, E. F. Lambin, E. Coast, A. Kariuki, I. Kikula, J. Kivelia, M. Said, S. Serneels and M. Thompson, 'Long-Term Changes in Serengeti-Mara Wildebeest and Land Cover: Pastoralism, Population, or Policies?', *Proceedings of the National Academy of Sciences of the United States of America*, XCVIII/22 (23 October 2001), p. 12546.

5 Jonathan and Angela Scot, *Big Cat Diary: Lion* (London, 2002), p. 12.

6 Laurence G. Frank, Rosie Woodroffe and Mordecai O. Ogada, 'People and Predators in Laikipia District, Kenya', in *People and Wildlife: Conflict or Coexistence?*, ed. Rosie Woodroffe, Simon Thirgood and Alan Rabinowitz (Cambridge, 2005), pp. 302–3.

7 Craig Packer, Dennis Ikanda, Bernard Kissui and Hadas Kushnir, 'Lion Attacks on Humans in Tanzania', *Nature*, CDXXXVI/18 (August 2005), p. 927.

8 Vasant K. Saberwal, James P. Gibbs, Ravi Chellam and A.J.T. Johnsingh, 'Lion-Human Conflict in the Gir Forest, India', *Conservation Biology*, VIII/2 (June 1994), p. 503.

9 Peter A. Lindsey, L.G. Frank, R. Alexander, A. Mathieson and S. S. Romañach, 'Trophy Hunting and Conservation in Africa: Problems and One Potential Solution', *Conservation Biology*, XXI/3 (2006), p. 880.

10 Ibid., p. 881.

11 Quoted by Geordie Torr, 'Living with Lions', *Geographical*, LXXVII/7 (2005), p. 63.

12 See Karyl Whitman, Anthony M. Starfield, Henley S. Quadling and Craig Packer, 'Sustainable Trophy Hunting of African Lions', *Nature*, CDXXVIII (2004), p. 175.

13 Lindsey et al., 'Trophy Hunting and Conservation in Africa', p. 882.

14 Michael Hopkin, 'Hunters Urged to Shoot Only Dark-nosed Lions', at www.bioedonline.org/news/news.cfm?art=794 [accessed 5 October 2009].

15 Divyabhanusinh, *Asia's Lions*, p. 199.

16 Derek Bousé, *Wildlife Films* (Philadelphia, PA, 2000), p. 23.

17 Ulf Eriksson and Örjan Johansson, 'Maasai Perceptions of Wildlife and People–Wildlife Interactions', *Minor Field Studies*, CCLX (Uppsala, 2004), p. 23.

18 Pieter Kat and Chris Harvey, *Prides: The Lions of Moremi* (Washington, DC, 2000), p. 29.

19 Barnabas Dickson, 'Global Regulation and Communal

Management', in *Endangered Species, Threatened Convention: The Past, Present, and Future of CITES*, ed. Jon Hutton and Barnabas Dickson (London, 2000), p. 176.

20 Packer et al., 'Lion Attacks on Humans in Tanzania', pp. 927–8.

21 Abolqasem Ferdowsi, *Shahnameh: The Persian Book of Kings*, trans. Dick Davis (New York, 2006), p. 48.

Select Bibliography

Adamik, Tamás, 'The Baptized Lion in the Acts of Paul', in *The Apocryphal Acts of Paul and Thecla*, ed. Jan N. Bremmer (Kampen, 1996), pp. 60–74

Adamson, Joy, *Born Free: A Lioness of Two Worlds* (London, 1960)

Aelian, *On the Characteristics of Animals*, trans. A. F. Scholfield, 3 vols (Cambridge, MA, 1959)

Aesop, *Fables of Aesop*, trans. S. A. Handford (London, 1954)

Alexander, Edward P., *Museum Masters: Their Museums and their Influence* (Walnut Creek, CA, 1983)

Allsen, Thomas T., *The Royal Hunt in Eurasian History* (Philadelphia, PA, 2006)

Alvey, Mark, 'The Cinema as Taxidermy: Carl Akeley and the Preservative Obsession', *Framework: The Journal of Cinema and Media*, XLVIII/1 (Spring 2007), pp. 23-45

Ashton, Dore and Denise Browne Hare, *Rosa Bonheur: A Life and a Legend* (New York, 1981)

Aulus Gellius, *Attic Nights*, 3 vols, trans. C. Rolfe (Cambridge, MA, 1984)

Barnes, Julian, *Talking it Over* (London, 1992)

Barnett, Ross, Nobuyuki Yamaguchi, Ian Barnes and Alan Cooper, 'The Origin, Current Diversity and Future Conservation of the Modern Lion (*Panthera leo*)', *Proceedings of the Royal Society of Biological Sciences*, CCLXXII (2006), pp. 2119–25

Barolsky, Paul, *Why Mona Lisa Smiles and Other Tales by Vasari* (University Park, PA, 1991)

Bartusis, Mark, Khalifa Ben Nasser and Angelikie E. Laiou, 'Days and Deeds of a Hesychast Saint: A Translation of the Greek Life of Saint Romylos', *Byzantine Studies / Etudes Byzantines*, ix/1 (1982), pp. 24–47

Bauer, H. and S. Van Der Merwe, 'Inventory of Free-Ranging Lions *Panthera leo* in Africa', *Oryx*, xxxviii/1 (2004), pp. 26–31

Baum, L. Frank, *The Wonderful Wizard of Oz* (London, 1987)

Bennett, Edward Turner, *The Tower Menagerie* (London, 1829)

Berger, John, *About Looking* (London, 1980)

Berry, H.H., 'Namibia's Seal-Eating Lions in Danger', *Cat News*, xiv (1991), p. 10

——, 'Last Skeleton Coast Lions Killed by Farmers', *Cat News*, xv (1991), pp. 7–8

Bostock, Frank C., *The Training of Wild Animals* (London and New York, 1903)

Bostock, Stephen St C., *Zoo and Animal Rights: The Ethics of Keeping Animals* (New York, 1993)

Bourne, Patricia, *Thank You, I Prefer Lions* (London, 1956)

Bousé, Derek, *Wildlife Films* (Philadelphia, pa, 2000)

Budge, Ernest A. Wallis, *Tutankhamen: Amenism, Atenism and Egyptian Monotheism* (London, 1923)

Bull, Bartle, *Safari: A Chronicle of Adventure* (London, 1988)

Burkhardt Jr, Richard W., 'A Man and His Menagerie', *Natural History*, cx/1 (February 2001), pp. 62–9

Cameron, Kenneth M., *Into Africa: The Story of the East African Safari* (London, 1990)

Carroll, Lewis, *Through the Looking-Glass and What Alice Found There* (London, 1948)

Cass, Caroline, *Joy Adamson: Behind The Mask*, 2nd edn (London, 2000)

Chauvet, Jean-Marie, Eliette Brunel Deschamps and Christian Hillaire, *Dawn of Art: The Chauvet Cave* (New York, 1996)

Clottes, Jean, and Marc Azéma, *Les félins de la grotte Chauvet* (Paris, 2005).

Clubb, Ros and Georgia Mason, 'Captivity Effects on Wide-Ranging Carnivores', *Nature*, cdxxv (2 October 2003), pp. 473–4

Conard, Nicholas J., 'Palaeolithic Ivory Sculptures from Southwestern

Germany and the Origins of Figurative Art', *Nature*, CDXXVI, 18/25 (December 2003), pp. 830–32

Currant, A. and R. Jacobi, 'Human Presence and Absence in Britain During the Early Part of the Late Pleistocene', in *Le Dernier Interglaciaire et les occupations du Paléolithique moyen*, ed. A. Tuffreau and W. Roebroeks (Lille, 2002), pp. 105–13

Dickson, Barnabas, 'Global Regulation and Communal Management', in *Endangered Species, Threatened Convention: The Past, Present, and Future of CITES*, ed. Jon Hutton and Barnabas Dickson (London, 2000), pp. 161–77

Divyabhanusinh, *The Story of Asia's Lions* (Mumbai, 2005)

Dobson, Jessie, 'John Hunter's Animals', *Journal of the History of Medicine*, XVII (1962), pp. 479–86

Donald, Diana, *Picturing Animals in Britain, 1750–1850* (New Haven, CT, and London, 2007)

Duffy, Stephen, 'Landseer and the Lion-Tamer: The 'Portrait of Mr Van Amburgh' at Yale', *British Art Journal*, III/3 (2002), pp. 25–35

Eastman, George, *Chronicles of an African Trip* (Rochester, NY, 1927)

Ebensperger, Luis A. 'Strategies and Counterstrategies to Infanticide in Mammals', *Biological Reviews*, LXXIII (1998), pp. 321–46

Eriksson, Ulf, and Örjan Johansson, 'Maasai Perceptions of Wildlife and People–Wildlife Interactions', *Minor Field Studies*, CCLX (Uppsala, 2004)

Estes, Richard Despard, *The Behaviour Guide to African Mammals* (Berkeley, CA, and London, 1991)

Ferdowsi, Abolqasem, *Shanameh: The Persian Book of Kings*, trans. Dick Davis (New York, 2006)

Findly, Ellison B., 'Jahangir's Vow of Non-Violence', *Journal of the American Oriental Society*, CVII/2 (1987), pp. 245–56

Frank, Laurence G., Rosie Woodroffe and Mordecai O. Ogada, 'People and Predators in Laikipia District, Kenya', in *People and Wildlife: Conflict or Coexistence?* ed. Rosie Woodroffe, Simon Thirgood and Alan Rabinowitz (Cambridge, 2005)

Franks, J. W., A. J. Sutcliffe, M. P. Kerney and G. R. Coope, 'Haunt of Elephant and Rhinoceros: The Trafalgar Square of 100,000 Years

Ago – New Discoveries', *The Illustrated London News* (14 June 1958), pp. 1011–13

Funston, Paul J., Michael G. L. Mills, Philip R. K. Richardson and Albert S. van Jaarsveld, 'Reduced Dispersal and Opportunistic Territory Acquisition in Male Lions (*Panthera leo*)', *Journal of Zoology*, CCLIX (2003), pp. 131–42

Gerald of Wales, *The History and Topography of Ireland*, trans. John J. O'Meara, (London, 1982)

Gittleman, John L., ed., *Carnivore Behavior, Ecology, and Evolution* (Ithaca, NY, 1989)

Grinnell, Jon, and Karen McComb, 'Maternal Grouping as a Defense Against Infanticide by Males: Evidence from Field Playback Experiments on African Lions', *Behavioral Ecology*, VII/1 (1996), pp. 55–9

——, 'Roaring and Social Communication in African Lions: The Limitations Imposed by Listeners', *Animal Behaviour*, LXII (2001), pp. 93–8

Guggisberg, C.A.W., *Simba: The Life of the Lion* (Cape Town, 1961)

Guillim, John, *A Display of Heraldrie* (London, 1610)

Guthrie, R. Dale, *Frozen Fauna of the Mammoth Steppe: The Story of Blue Babe* (Chicago, 1989)

Hahn, Daniel, *The Tower Menagerie: The Amazing True Story of the Royal Collection of Wild Beasts* (London, 2003)

Hemingway, Ernest, *The Complete Short Stories of Ernest Hemingway: The Finca Vigia Edition* (New York, 1987)

——, *True at First Light* (London, 2004)

Herne, Brian, *White Hunters: The Golden Age of African Safaris* (New York, 1999)

Holloway, Steven W., 'Biblical Assyria and Other Anxieties in the British Empire', *Journal of Religion and Society*, III (2001), pp. 1–19

Homewood, K., E. F. Lambin, E. Coast, A. Kariuki, I. Kikula, J. Kivelia, M. Said, S. Serneels and M. Thompson, 'Long-term Changes in Serengeti-Mara Wildebeest and Land Cover: Pastoralism, Population, or Policies? *Proceedings of the National Academy of Sciences of the United States of America*, XCVIII/22 (23 October 2001), pp. 12544–9

Höner, Oliver P., Bettina Wachter, Marion L. East and Heribert Hofer,

'The Responses of Spotted Hyenas to Long-Term Changes in Prey Populations: Functional Response and Interspecific Kleptoparasitism', *Journal of Animal Ecology*, LXXI (2002), pp. 236–46

Jackson, Deirdre, *Marvellous to Behold: Miracles in Medieval Manuscripts* (London, 2007)

Jackson, Sir Frederick, *Early Days in East Africa* (London, 1930)

Johnson, William, *The Rose-Tinted Menagerie* (London, 1990)

Johnston, William R., and Simon Kelly, *Untamed: The Art of Antoine-Louis Barye* (Baltimore, MD, 2006)

Joubert, Dereck, 'Hunting Behaviour of Lions (*Panthera leo*) on Elephants (*Loxodonta africana*) in the Chobe National Park, Botswana', *African Journal of Ecology*, XLIV (2006), pp. 279–81

Joubert, Dereck and Beverly Joubert, *Hunting with the Moon* (Washington, DC, 1997)

Kat, Pieter and Chris Harvey, *Prides: The Lions of Moremi* (Washington, DC, 2000)

Keeney, Bradford, ed., *Ropes to God: Experiencing the Bushman Spiritual Universe* (Philadelphia, PA, 2003).

Kerr, Alex, *No Bars Between* (London, 1957)

Kisling, Vernon N., ed., *Zoo and Aquarium History: Ancient Animal Collections to Zoological Gardens* (Boca Raton, FL, 2000)

Kliman, Eve Twose, 'Delacroix's Lions and Tigers: A Link between Man and Nature', *Art Bulletin*, LXIV/3 (1982), pp. 446–66

Klingender, Francis, *Animals in Art and Thought to the End of the Middle Ages* (London, 1971)

Klumpke, Anna, *Rosa Bonheur: The Artist's (Auto)biography* (Ann Arbor, MI, 2001)

Kruuk, Hans, *Hunter and Hunted: Relationships Between Carnivores and People* (Cambridge, 2002)

Lacy, Norris, J., ed., *Lancelot–Grail: The Old French Arthurian Vulgate and Post-Vulgate in Translation*, vol. III, trans. Roberta L. Krueger, William W. Kibler and Carleton W. Carroll (New York and London, 1995)

Layard, Austen Henry, *Early Adventures in Persia, Susiana, and Babylonia*, 2 vols (Piscataway, NJ, 2003)

Leutemann, Heinrich, *Lebensbeschreibung des Thierhändlers Carl Hagenbeck* (Hamburg, 1887)

Lindsey, Peter A., L. G. Frank, R. Alexander, A. Mathieson and S. S. Romañach, 'Trophy Hunting and Conservation in Africa: Problems and One Potential Solution', *Conservation Biology*, XXI/3 (2006), pp. 880–83

Lonsdale, Steven H., *Creatures of Speech: Lion, Herding and Hunting Similes in the Iliad* (Stuttgart, 1990)

MacKenzie, J. M., *The Empire of Nature: Hunting, Conservation and British Imperialism* (Manchester, 1988)

Mandeville, Bernard, *The Fable of the Bees, or Private Vices, Publick Benefits*, ed. Irwin Primer (New York, 1962)

Margulis, Susan W., Catalina Hoyos and Meegan Anderson, 'Effect of Felid Activity on Zoo Visitor Interest', *Zoo Biology*, XXII (2003), pp. 587–99

McComb, Karen, Craig Packer and Anne Pusey, 'Roaring and Numerical Assessment in Contests Between Groups of Female Lions', *Animal Behaviour*, XLVII (1994), pp. 379–87

Mitman, Gregg, *Reel Nature: America's Romance with Wildlife on Films* (Cambridge, MA, 1999)

Moore, Wendy, *The Knife Man: The Extraordinary Life and Times of John Hunter, Father of Modern Surgery* (London, 2005)

Mras, George P., 'A Crouching Royal Tiger by Delacroix', *Record of the Art Museum, Princeton University*, XXI/1 (1962), pp. 16–24

Murombedzi, James C., 'Pre-Colonial and Colonial Conservation Practices in Southern Africa and their Legacy Today', *IUCN, The World Conservation Union* (2003), pp. 1–21

Nelson, Robert H. 'Environmental Colonialism: "Saving" Africa from Africans', *Independent Review*, VIII/1 (Summer 2003), pp. 65–86

Nizami, *Layla and Majnun*, ed. and trans. Colin Turner (London, 1997)

Norton, Lucy, trans., *The Journal of Eugène Delacroix* (London, 1951).

O'Brien, Stephen J., *Tears of the Cheetah: The Genetic Secrets of Our Animal Ancestors* (New York, 2003)

Packer, Craig, *Into Africa* (Chicago, 1994)

Packer, Craig, Dennis Ikanda, Bernard Kissui and Hadas Kushnir,

'Lion Attacks on Humans in Tanzania', *Nature*, CDXXXVI/18 (August 2005), pp. 927–8

Patterson, Bruce D., *The Lions of Tsavo: Exploring the Legacy of Africa's Notorious Man-Eaters* (New York and London, 2004)

——, 'On the Nature and Significance of Variability in Lions (*Panthera leo*)', *Evolutionary Biology*, XXXIV (2007), pp. 55–60

Pease, Alfred Edward, *The Book of the Lion* (London, 1913)

Peters, G., 'Purring and Similar Vocalizations in Mammals', *Mammal Review*, XXXII/4 (2002), pp. 245–71

Pliny the Elder, *Natural History*, trans. H. Rackham, W.H.S. Jones and D. E. Eichholz, 10 vols (London, 1938–62)

Rainsford, William, S. *The Land of the Lion* (London, 1909)

Randolph, Adrian W. B., 'Il Marzocco: Lionizing the Florentine State', in *Coming About... A Festschrift for John Shearman*, ed. Lars R. Jones and Louisa C. Matthews (Cambridge, MA, 2001), pp. 11–18

Reade, Julian, *Assyrian Sculpture* (London, 1983)

Renard, John, *Friends of God: Islamic Images of Piety, Commitment, and Servanthood* (Berkeley and Los Angeles, CA, 2008)

Richmond, Velma Bourgeois, *The Legend of Guy of Warwick* (New York and London, 1996)

Ross, Doran H., 'The Heraldic Lion in Akan Art: A Study of Motif Assimilation in Southern Ghana', *Metropolitan Museum Journal*, XVI (1981), pp. 165–80

Rothfels, Nigel, *Savages and Beasts: The Birth of the Modern Zoo* (Baltimore, MD, and London, 2002)

Rudnai, Judith A., *The Social Life of the Lion: A Study of the Behaviour of Wild Lions in the Nairobi National Park, Kenya* (Lancaster, 1973)

Saberwal, Vasant K., James P. Gibbs, Ravi Chellam and A.J.T. Johnsingh, 'Lion–Human Conflict in the Gir Forest, India', *Conservation Biology*, VIII/2 (June 1994), pp. 501–7

Scarfi, Bianca Maria, ed., *The Lion of Venice: Studies and Research on the Bronze Statue in the Piazzetta* (Venice, 1990)

Schaller, George B., *The Serengeti Lion: A Study of Predator–Prey Relations* (Chicago and London, 1972)

——, *Serengeti: A Kingdom of Predators* (New York, 1972)

——, *Golden Shadows, Flying Hooves* (London, 1973)

Schapera, I., 'A Native Lion Hunt in the Kalahari Desert', *Man*, XXXII (December 1932), pp. 278–82

Scot, Jonathan and Angela, *Big Cat Diary: Lion* (London, 2002)

Scribner, Charles, *Peter Paul Rubens* (London, 1992)

Scull, Guy H., 'Lassoing Wild Animals in Africa', *Everybody's Magazine*, XXIII/3–5 (1910), pp. 309–22, 526–38, 609–21

Selous, Frederick Courteney, *A Hunter's Wanderings in Africa: Being a Narrative of Nine Years Spent amongst the Game of the Far Interior of South Africa* (London, 1881)

Shaw, George Bernard, *Androcles and the Lion* (Harmondsworth, 1949)

Shaw, Margaret Renée Bryers, trans., *Joinville and Villehardouin: Chronicles of the Crusades* (London, 1963)

Singh, H.S., 'Population Dynamics, Group Structure and Natural Dispersal of the Asiatic lion (*Panthera leo persica*)', *Journal of the Bombay Natural History Society*, XCIV/1 (1997), pp. 65–70

Sotnikova, Marina, and Pavel Nikolskiy, 'Systematic Position of the Cave Lion *Panthera spelaea* (Goldfuss) Based on Cranial and Dental Characters', *Quaternary International*, CXLII– CXLIII (2006), pp. 218–28

Stander, P. E., 'Cooperative Hunting in Lions: The Role of the Individual', *Behavioural Ecology and Sociobiology*, XXIX (1992), pp. 445–54

Steinhart, Edward I., *Black Poachers, White Hunters: A Social History of Hunting in Colonial Kenya* (Oxford, 2006)

Stokes, John, '"Lion Griefs": The Wild Animal Act as Theatre', *New Theatre Quarterly*, XX (2004), pp. 138–54

Storey, William K., 'Big Cats and Imperialism: Lion and Tiger Hunting in Kenya and Northern India, 1898–1930', *Journal of World History*, II/2 (Fall 1991), pp. 135–73

Strawn, Brent A., *What Is Stronger than a Lion? Leonine Image and Metaphor in the Hebrew Bible and the Ancient Near East* (Göttingen, 2005)

Strong, John S., 'The Legend of the Lion-Roarer: A Study of the Buddhist Arhat Pindola Bharadvaja', *Numen*, XXVI/1 (1979), pp. 50–88

Tanavoli, Parviz, *Lion Rugs: The Lion in the Art and Culture of Iran* (Basel, 1985)

Torr, Geordie, 'Living with Lions', *Geographical*, LXXVII/7 (2005), pp. 60–64

Trinkel, Martina and Gerald Kastberger, 'Competitive Interactions between Spotted Hyenas and Lions in the Etosha National Park, Namibia', *African Journal of Ecology*, XLIII (2005), pp. 220–24

Turner, Alan and Mauricio Antón, *The Big Cats and their Fossil Relatives: An Illustrated Guide to their Evolution and Natural History* (New York, 1997)

Unterman, Alan, *Dictionary of Jewish Lore and Legend* (London, 1997)

Waddell, Helen, *Beasts and Saints* (London, 1934)

Walker, Clive, *Signs of the Wild: A Field Guide to the Tracks and Signs of the Mammals of Southern Africa*, 5th edn (Cape Town, 1996)

Weissengruber, G. E., G. Forstenpointner, G. Peters, A. Kübber-Heiss, and W. T. Fitch, 'Hyoid Apparatus and Pharynx in the Lion (*Panthera leo*), Jaguar (*Panthera onca*), Tiger (*Panthera tigris*), Cheetah (*Acinonyx jubatus*) and Domestic Cat (*Felis silvestris* f. *catus*)', *Journal of Anatomy*, CCI/3 (2002), pp. 195–209

Werdelin, Lars, and Margaret E. Lewis 'Plio-Pleistocene Carnivora of eastern Africa: Species Richness and Turnover Patterns', *Zoological Journal of the Linnean Society*, CXLIV/2 (2005), pp. 121–44

West, Peyton M. and Craig Packer, 'Sexual Selection, Temperature, and the Lion's Mane', *Science*, 297 (23 August 2002), pp. 1339–43

White, T. H., *The Bestiary: A Book of Beasts* (New York, 1960)

Whitman, Karyl, Anthony M. Starfield, Henley S. Quadling, and Craig Packer, 'Sustainable Trophy Hunting of African Lions', *Nature*, CDXXVIII (2004), pp. 175–78

Wienholt, Arnold, *The Story of a Lion Hunt* (London and New York, 1922)

Wittgenstein, Ludwig, *Philosophische Untersuchungen* (*Philosophical Investigations*), trans. G.E.M. Anscombe (Oxford, 1953)

Yamaguchi, Nobuyuki, Alan Cooper, Lars Werdelin and David W. Macdonald, 'Evolution of the Mane and Group-living in the Lion (*Panthera leo*): A Review', *Journal of the Zoological Society of London*, CCLXIII (2004), pp. 329–42

Associations and Websites

AFRICAN LION WORKING GROUP
www.african-lion.org

THE ASIATIC LION INFORMATION CENTRE
www.asiatic-lion.org

BIBLIOTHÈQUE NATIONALE DE FRANCE: MEDIEVAL BESTIARY
expositions.bnf.fr/bestiaire/expo/

BIG CATS INITIATIVE
www.nationalgeographic.com/bigcats

BORN FREE FOUNDATION
www.bornfree.org.uk

THE BRITISH LIBRARY DIGITAL CATALOGUE OF ILLUMINATED
MANUSCRIPTS
www.bl.uk/catalogues/illuminatedmanuscripts/welcome.htm

THE ELSA CONSERVATION TRUST
www.elsatrust.org

LION CONSERVATION FUND
www.lionconservationfund.org

THE LION RESEARCH CENTER
www.lionresearch.org

LIVING WITH LIONS
www.lionconservation.org

PREDATOR CONSERVATION TRUST
www.predatorconservation.com

Acknowledgements

It is a pleasure to thank those who helped me while I was writing this book. I spent a memorable morning with Andy Currant at the Natural History Museum, London. I am grateful to him for showing me drawer upon drawer of lion fossils and regaling me with stories of their discovery. Malcolm Harman gave me a guided tour of the collection at Quex Museum, which includes the Burberry jacket the Major was wearing when he was mauled by a lion in the Congo (the lion itself is also on display). At the Grant Museum of Zoology (UCL), which holds 62,000 specimens, I examined a range of lion skulls with Mark Carnall. Other specimens of *Panthera leo* are on exhibition at the Hunterian Museum at the Royal College of Surgeons of England, and countless leonine artefacts are held by the British Museum; I drew inspiration from both collections.

Visits to London Zoo were enhanced by the birth of two Asian lion cubs on 8 June 2009; I am grateful to parents Lucifer and Abi for their impeccable timing. Special thanks to my colleagues at the British Library, including Nicolas Bell, Sarah Biggs, Claire Breay, Justin Clegg, Kathleen Doyle, Juan Garcés, Julian Harrison, Chris Lee, Scot McKendrick, John Rhatigan, Barry Taylor, and the staff of the Manuscripts Reading Room. I am also indebted to Alixe Bovey, Jean Clottes, Peter Coates, Jill Cook, Catherine Haill, Gayle Haugen, Garnet and Gayle Jackson, Daniel Lacy, John Lowden, Laura Nuvoloni, John Osborne, Ann Sylph, Kurt Wehrberger, Kurt and Charlotte Weinberg, and Catherine Yvard. It has been a privilege to work with Michael Leaman, Martha Jay and Harry Gilonis at Reaktion, and with Jonathan

Burt, editor of the Animal series. Finally, I thank the Scouloudi Foundation, in association with the Institute of Historical Research, University of London, for awarding me a grant to defray the cost of illustrations.

Photo Acknowledgements

The author and publishers wish to express their thanks to the below sources of illustrative material and/or permission to reproduce it. (Some sources uncredited in the captions for reasons of brevity are also given below.)

Photo Alinari/Rex Features: p. 98; photo © Denis Ananiadis/2009 iStock International Inc.: p. 32; collection of the author: pp. 58, 64, 91, 94, 95; from L. Frank Baum, *The Wonderful Wizard of Oz* (Chicago, 1900): p. 136; from Edward Turner Bennett, The *Tower Menagerie* (London, 1829): pp. 20, 149; Bibliothèque nationale de France, Paris (photos © Bibliothèque nationale de France): pp. 104 (MS Français 113, f. 116v), 129 (MS Latin 8865, f. 43), 145 (MS Français 245, f. 119v); from Frank C. Bostock, *The Training of Wild Animals* (London and New York, 1903): pp. 67, 68; The British Library, London (photos © British Library Board): pp. 73 (Additional MS 5600, f. 419v), 74 (Royal MS 13 B VIII, f. 19v), 105 (Harley MS 4903, f. 135), 107 (Yates Thompson MS 13, ff. 9v and 12), 109 (Additional MS 23770, f. 13v), 110 (Yates Thompson MS 8, f. 4); 127 (Royal 12 C XIX, f. 6); 134 (Royal 12 F XIII, f. 5v), 147 (Or. 2265, f. 166); 152 (Additional MS 15153, f. 95v); The British Museum, London (photos © Trustees of the British Museum): pp. 78, 79, 80, 82, 85, 97, 99, 101, 108, 111, 113, 118, 125, 126, 131, 135, 140, 141, 143, 146, 148, 159, 160; from Lewis Carroll, *Through the Looking-Glass and what Alice Found There* (London, 1948): p. 138; reproduced courtesy of J. Clottes: p. 12; photo © Sue Colvil/2009 iStock International Inc.: p. 132; from William Darton, *Present for a Little Boy* (London, 1798): p. 76; photo ©

Ricardo De Mattos/2009 iStock International Inc.: p. 194; photos © Markus Divis/2009 iStock International Inc.: pp. 23, 51; from George Eastman, *Chronicles of an African Trip* (Rochester, NY, 1927): pp. 180, 182; photo © Holger Ehlers/2009 iStock International Inc.: p. 29; reproduced courtesy of the Elsa Conservation Trust: p. 55; photo © Giorgio Fochesato/2009 iStock International Inc.: p. 128; photo © Dirk Freder/2009 iStock International Inc.: p. 24; photos © David T. Gomez/ 2009 iStock International Inc.: pp. 36, 37; photo © Albert Harlingue/ Roger-Viollet, courtesy Rex Features: p. 81; photo © Jonathan Heger/ 2009 iStock International Inc.: p. 192; Ernest Hemingway Collection, John F. Kennedy Library, Boston, MA: p. 177; Hunterian Museum, London, photo © The Royal College of Surgeons of England: p. 84; photos © Britta Kasholm-Tengve/2009 iStock International Inc.: pp. 10, 164; from Robert Koldewey, *The Excavations at Babylon* (London, 1914): p. 112 (bottom); photo © Daniel Lacy: p. 121; photo © NirLasman/ 2009 iStock International Inc.: p. 116; photo © Ma Liang/2009 iStock International Inc.: p. 120; Library of Congress, Washington, DC (Prints and Photographs Division): pp. 61, 66, 88, 169, 172; photo © Klaas Lingbeek-van Kranen/2009 iStock International Inc.: p. 26; from David Livingstone, *Missionary Travels and Researches in South Africa* (London, 1857): p. 162; photos © Peter Malsbury/2009 iStock International Inc.: pp. 42, 43, 48, 190; photo © Manfredxy/2009 iStock International Inc.: p. 6; photo © Peter Miller/2009 iStock International Inc.: p. 46; from Alfred E. Pease, *Travel and Sport in Africa* (London, 1902): p. 155; photo © Ogen Perry/2009 iStock International Inc.: p. 30; photo © Graeme Purdy/2009 iStock International Inc.: p. 35; photo Baron Nicholas de Racozcy: p. 71; photo © Brian Raisbeck/2009 iStock International Inc.: p. 50; Rex Features: p. 57; photo Andy Rouse/Rex Features: p. 38; The Royal Collection © 2009, Her Majesty Queen Elizabeth II: pp. 59, 157; Royal Geographical Society Archive, London: p. 174; photos © Pradeep Kumar Saxena/2009 iStock International Inc.: pp. 17, 124, 191; photo Science and Society Picture Library: p. 90; from Frederick Courteney Selous, *A Hunter's Wanderings in Africa* (London, 1895): pp. 165, 167; photo © Wolfgang Steiner/2009 iStock International Inc.: p. 186; photo Thomas Stephan (© Ulmer Musuem,

Ulm): p. 15; photo © Gert Vrey/2009 iStock International Inc.: p. 27; Victoria and Albert Museum, London (© V&A Images - all rights reserved): pp. 71 (Theatre Collections), 87, 112 (top), 114, 123, 161; photo © Mark Wilson/2009 iStock International Inc.: p. 25; photo © Pauline Wilson/2009 iStock International Inc.: p. 187; photo © Kirk Winslow/2009 iStock International Inc.: p. 34; photo © Kate Womack/ 2009 iStock International Inc.: p. 130; photo © David Youldon/2009 iStock International Inc.: p. 40.

Index